More modern chemical techniques

Written by Ralph Levinson

School Teacher Fellow 1995–1996

CR
543
LEV

36064147X

ROYAL SOCIETY OF CHEMISTRY

More modern chemical techniques

Written by Ralph Levinson

Edited by Martyn Berry, John Johnston, Colin Osborne and Maria Pack

Designed by Imogen Bertin and Sara Roberts

Published by The Royal Society of Chemistry

Printed by The Royal Society of Chemistry

Copyright © The Royal Society of Chemistry 2001

Apart from any fair dealing for the purposes of research or private study, or criticism or review, as permitted under the UK Copyright Designs and Patents Act, 1988, this publication may not be reproduced, stored, or transmitted, in any form or by any means, without the prior permission in writing of the publishers, or in the case of reprographic reproduction, only in accordance with the terms of the licences issued by the Copyright Licensing Agency in the UK, or in accordance with the terms of licences issued by the appropriate Reproduction Rights Organisation outside the UK. Enquiries concerning reproduction outside the terms stated here should be sent to the Royal Society of Chemistry at the London address printed on this page.

Notice to all UK Educational Institutions. The material in this book may be reproduced by photocopying for distribution and use by students within the purchasing institution providing no more than 50% of the work is reproduced in this way for any one purpose. Tutors wishing to reproduce material beyond this limit or to reproduce the work by other means such as electronic should first seek the permission of the Society.

While every effort has been made to contact owners of copyright material we apologise to any copyright holders whose rights we may have unwittingly infringed.

For further information on other educational activities undertaken by the Royal Society of Chemistry write to:

Education Department
Royal Society of Chemistry
Burlington house
Piccadilly
London W1J 0BA

Information on other Royal Society of Chemistry activities can be found on its websites:
www.rsc.org
www.chemsoc.org
www.chemsoc.org/LearnNet contains resources for teachers and students from around the world.

ISBN 0–85404–929–0

British Library Cataloguing in Publication Data.

A catalogue for this book is available from the British Library.

RS•C

Foreword

Analytical techniques are powerful tools in a chemist's armoury. Spectroscopic data and chemical information are used routinely in laboratories to follow a chemical reaction or elucidate a chemical structure. However, the sophistication of the analytical techniques used changes rapidly, hence the routinely used method of today can all too readily be superseded by the new technology of tomorrow. This book identifies some applications of the important chemical techniques in use today that are less well known in schools and colleges and which illustrate how chemistry is using state-of-the-art technology to push back the frontiers of the subject.

Professor Tony Ledwith CBE PhD
DSC CChem FRSC FRS
President, The Royal Society of Chemistry

RS•C

Contents

RS•C

Introduction

Every aspect of life in a modern technological society depends on the analysis of chemicals. Food, fuel, construction materials, medicines, industrial effluents are among the many materials that are analysed to determine their chemical contents; indeed it is difficult to think of a single product or by-product that is not sampled and chemically analysed during its life-time.

The Royal Society of Chemistry addresses the teaching of modern analytical techniques through the many courses it runs for teachers in conjunction with industry and university chemistry departments as well as a number of publications, a video and CD-ROM. The video, Modern Chemical Techniques, describes techniques which are widely used in industry and research. Between them, this book and its predecessor, Modern Chemical Techniques, describes all the techniques in the video and many others.

I spent a fascinating year as teacher fellow at the RSC researching and writing this book, its primary aim being to inform and update teachers on analytical techniques commonly used in industry, research and other analytical laboratories. A number of these techniques may be unfamiliar to teachers and students in schools though they are very versatile and can be made to determine minute amounts of material. Inductively Coupled Plasma (ICP) techniques, for example, predominate in the detection and quantification of metals. When an ICP torch is coupled to a mass spectrometer, metals can be detected down to parts per trillion; in other words, one gram of metal in a swimming pool of water. Capillary electrophoresis is a formidable technique which is capable of separating quantities of material in the order of atto (10^{-18}) moles. Given the level of sophistication, expense and automation of the instruments, it is hardly surprising they are not found in schools but their chemistries can be explained even at pre-GCSE level, and some of their applications reveal new and amazing possibilities.

The techniques in this book cover qualitative and quantitative determinations of elements, separation systems and the structural elucidation of materials. In each chapter there is a description of the principles of the technique and the instrument, its state-of-the-art technology and applications. Even the most sophisticated data produced by expensive and highly automated instruments is only as good as the sample being analysed. If the sample is contaminated and/or unrepresentative of the batch from which it is taken, then the whole analytical procedure is worthless. Where relevant, the chapter contains a section on sample preparation, instrument calibration and carrying out an analysis. A separate section describes general approaches to sampling.

I hope this book will offer a varied experience to its readers. Those who want a basic understanding of a technique should derive benefit from it, others will be able to delve more deeply. The boxes indicate parts which may be skipped by the general reader. Above all, it aims to give a sense that chemical analysis is about solving challenging problems using some wonderful tools.

Ralph Levinson

RS•C

Acknowledgements

Many people made significant contributions to this book.

Jocelyn Brown, Sandra Bell and Mary Walsh of the Education Department at Burlington House provided secretarial help which was always carried out in a friendly and selfless way.

Colin Whittaker at the Port Sunlight laboratory of Unilever gave me a great deal of his time explaining the challenges of atomic spectrometry and responding to drafts promptly and critically.

Joel Aliphon of Beckman introduced me to the spectacular technology of capillary electrophoresis. He posed some analytical problems for me to try and solve, generously giving me the opportunity to use Beckman equipment.

Don Bradley hosted my informative day at the Analytical Geochemistry department of the British Geological Survey (BGS) which gave an Earth Science context to most of the techniques described in this book. Don commented on the draft material and gave much helpful advice as did Vicki Hards and Simon Chenery also of the BGS. Vicki and Simon demonstrated X-ray diffraction and ICP-MS respectively, considerably enhancing my knowledge of these techniques.

John Theobald of Fisons spent a morning updating my knowledge of microelemental analysis. Malcolm Rose of the Open University explained the background of capillary electrophoresis and made detailed comments on my draft chapter. David Jones arranged my day's visit to Unilever's Colworth laboratory, which was very valuable, and commented in detail on the material I wrote.

Keith Marshall arranged my visit to the Laboratory of the Government Chemist (LGC), and promptly responded to any requests I made for information. John Pearce, Tony Stephenson and Indu Patel at the LGC all provided me with a lot of background information.

Two people played a major part in the whole project. John Reed organised my lengthy visits to Unilever's Port Sunlight laboratory. My visits to Port Sunlight were productive and enjoyable. Anything I requested from John seemed to materialise as soon as I asked for it.

Neville Reed at the RSC supervised the project, providing great support and encouragement. Neville was always available to discuss ideas and responded with impressive speed to lengthy drafts that I left on his desk.

The following people gave me their time and expertise:

John Birmingham	Unilever, Port Sunlight
Allan Clarke	Unilever, Colworth
Kevin Dilley	Unilever, Colworth
Ben Fairman	LGC
David Hartshorne	Cellmark Diagnostics
Andy Homan	Unilever, Colworth
Dean Madden	National Centre for Biotechnology Education
Hemlata Pancholi	Astra, Charmworth
Scott Singleton	Unilever, Port Sunlight
Lesley Smart	The Open University
Ian Smith	Unilever, Port Sunlight
David Taylor	Astra, Charmworth
Chris Tier	Unilever

RS•C

I was fortunate in having critical readers who provided me with constructive comments and in some cases, new information. Victoria Bearman, Mandy Trezise, Jean Pocock, Gene Robinson, Rose Hoban, Judy Machin, Professor David Snaith, Greg Herdman, Dr R.E.E Hill and Paul Board responded to most of my work with humour, analytical depth and frequently updated my knowledge. Other helpful readers were:

Prof A.T. Andrews
Linda Ault
Dr Nigel Barber
Philip Bearman
Victoria Bearman
Martyn Berry
Paul Board
J.T. Bodiam
Dr C. Bullock
Martin Carr
Michael J. Carr
Dr B. Crowther
P.E. Curry
Len Davenport
Dr D.E.P. Davies
J. Davies
Dr M. Drennan
Hugh Dunlop
David Everett
Ben Faust
Graham Fisher
Neil Fortey
Alan Furse
J. Garwell
A.R. Gouldson
Dr Eddie Grimble
John Haddon
L.A. Hambrook
G.A. Herdman
Dr R.E. Hill
Rose Hoban
Dr G.M. Hornby
Anne Hurworth
Mark Ingham
David Jones
Graham Kingsley
Paul Kodur
C.D. Law
James McAdam
Judy Machin
A. Mackenzie
M. Mahes
Tim Meunier
Dr David Moore
Dr C. Rees
Dr M.E. Rose
Gene Robinson
Sherry Rutland

RS•C

Professor P.W. Scott
Dr Adrian Shaw
David Snaith
Jenny Spouge
M. Thomas
Mandy Trezise
Elspeth Vicars
Chris Wood
G. Woods

The Society is indebted to Martyn Berry for his work in editing the drafts and obtaining permissions and to Dr Liz Pritchard of the LGC for a wealth of helpful comments.

The chapter on Optical Microscopy was originally produced by Ben Faust, RSC Teacher Fellow 1991–1992, and has been edited by the author.

RS•C

RS•C

Atomic absorption spectrometry

Introduction

The interaction of light with a sample can be used to give qualitative and quantitative information about the component elements in the sample under investigation. In atomic spectrometry, a sample is atomised by heating. Elements can be identified and quantified either by the emission or absorption of specific wavelengths of light.

Flame tests illustrate broadly the principle of atomic spectrometry. When some potassium chloride crystals are scattered on the flame of a Bunsen burner, a lilac colouration demonstrates the presence of potassium ions. If more crystals are added the flame turns a deeper colour. These two observations tell us first, the identity of the metal ion and secondly, a rough comparison of the amounts of the ion in the flame.

The potassium ions are excited by the energy of the flame. Electrons are promoted to higher energy levels, and as they return to their original positions, radiation of specific wavelengths is emitted. For potassium the wavelengths correspond to a visible part of the electromagnetic spectrum. Increasing the amount of potassium ions in the flame increases the intensity of the emission.

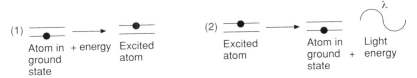

Figure 1 Schematic diagram of electron transition and light emission

When atoms absorb electromagnetic radiation the same amount of energy is emitted when they relax, that is, the electrons return to the ground state, the most stable electronic configuration for an element. This corresponds to the electrons occupying the orbitals with the lowest energy levels.

Flame tests are an example of atomic emission. The two main types of atomic spectrometry involve emission and absorption. In atomic absorption the atoms of an element absorb just enough radiant energy to promote electrons to a higher energy level. The amount of light absorbed is quantified and the results are processed to indicate the concentration of the element in a sample.

Atomic spectrometry has a wide variety of applications in industry, research, the forensic sciences and environmental monitoring. For example:

■ the permitted concentrations of heavy metals in water can be monitored;

■ the levels of phosphorus and antimony in flame-proofing materials have to be carefully maintained between lower and upper limits; and

■ determining trace metals in pottery shards – in archaeology – can be used to reveal ownership.

RS•C

Wavelengths and photons **Box 1**

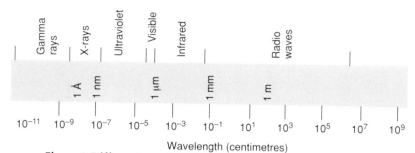

Figure 2 Different wavelengths in the electromagnetic spectrum

Light radiation generated from a source consists of an electromagnetic field oscillating periodically in the form of a sine wave.

Some waves have long wavelengths – *eg* radio waves can be 1000 m or more. The values of the wavelengths of X-rays and gamma rays are less than 10^{-7} m.

There is an inverse relationship between wavelength and frequency (the number of oscillations per second of the electromagnetic field). Radio waves have low frequencies whereas wavelengths in the ultraviolet and X-ray regions have high frequencies. The relationship between wavelength and frequency can be expressed in the form of an equation:

$$c = f\lambda$$

where c = velocity/ms^{-1}, f = frequency/s^{-1} or Hertz/Hz and λ = wavelength/m

The characterisation of light as waves is consistent with properties such as reflection, refraction and diffraction. Other properties of light such as the photoelectric effect can be better explained in terms of the particle theory of light, hence light of a particular frequency or wavelength is also associated with small packets of energy known as photons, as exhibited when detecting light in atomic absorption spectrometers. Higher frequency light has more energetic photons, so X-rays have a greater amount of energy associated with them than visible light or radio waves.

The energy of a photon is related to frequency (v) through Planck's constant:

$$E = hv$$

where E = energy (J) and h = Planck's constant (6.623 x 10^{-34} Js)

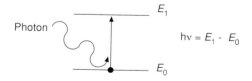

Figure 3 The interaction of a photon with an atom and the promotion of an electron

The energy levels in atoms are quantised, hence the electrons can only be promoted if the energy imparted to the atom corresponds exactly to the energy difference between the two levels. Before the promotion of an electron the atom is in the ground state. When photons with the correct energy interact with atoms in the ground state, the electrons are promoted from the lowest energy level to a higher energy level.

RS•C

The hydrogen spectrum   **Box 2**

The relationship between the quantum numbers of the energy levels and an emission spectrum can be illustrated by the hydrogen spectrum. Five distinct series of lines appear in the atomic emission spectrum of hydrogen.

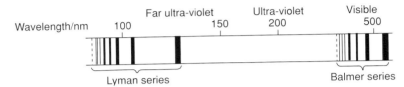

Figure 4 A hydrogen spectrum showing the Lyman and Balmer series

Each series of lines shows a similar pattern; the spacing between them narrows with decreasing wavelengths until the lines appear to converge. In the Lyman series, each line corresponds to an electron falling from a higher energy level to the ground state. The relationship between the wavelength of each line in the Lyman series and the electron transitions is expressed by the equation:

$$1/\lambda = R_\infty (1 - 1/n^2)$$

where

λ = wavelength of the electromagnetic radiation (m)
R_∞ = a constant known as the Rydberg constant = 1.097×10^5 cm^{-1}
n = a whole number > 1

For the Balmer and Paschen series the transitions are *not* to the ground state.

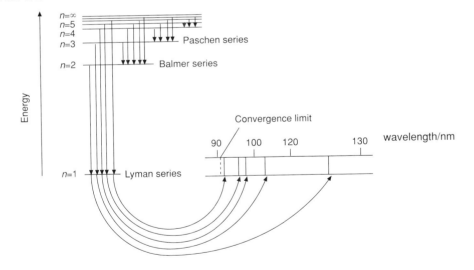

Figure 5 Relating Lyman series to spectral lines

The basis of atomic absorption spectrometry (AAS) is measuring the amount of light absorbed by a cloud of atoms when electrons are promoted from the ground state to a higher energy level.

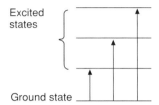

Figure 6 Electronic transitions in absorption

Atomic absorption spectrometry (AAS) is used to measure the concentrations of elements down to parts per billion, ppb (μg dm^{-3}), ($10^{-6} g$ dm^{-3}). A sample is heated in a flame to form dissociated atoms which absorb light from a source.

There is a direct relationship between the amount of light absorbed and the concentration of the element in the light path. The substance being analysed is commonly referred to as the analyte.

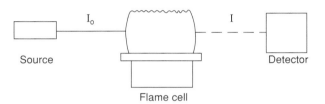

Figure 7 The change in light intensity in flame atomic absorption

Light of intensity, I_o, is passed into the flame cell containing the atoms in their ground state. Depending on the number of atoms in the flame cell the intensity of the light is reduced to a value, I. For mathematical purposes this fraction of light absorbed is converted to a value, absorbance (A).

$A = \log (I_o/I)$

There is a linear relationship between the absorbance (A) and the concentration (c) of the element. This is expressed through Beer's Law:

$A = \log (I_o/I) = \varepsilon c l$

where:

ε = the molar absorptivity coefficient and is constant for a particular element at a particular wavelength

c = the concentration of analyte in the sample in parts per million (ppm) or mg dm^{-3}

l = the path length through the sample cell (m).

There is a straight line relationship between absorbance (A) and concentration (c) that becomes non-linear at high concentrations.

RS•C

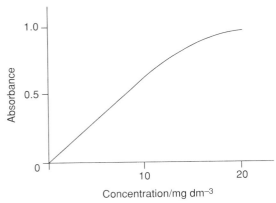

Figure 8 The relationship between absorbance (A) and concentration in flame atomic absorption

This relationship is used in calibrating the instrument. The absorbance values of standards of accurately known concentrations are measured. A graph is drawn and concentrations of unknowns can then be calculated from their absorbance values.

Instrumentation

There are three main components to all atomic absorption spectrometers. They are:

■ a light source;

■ an atom cell containing ground state atoms of the sample; and

■ a means of detecting specific wavelengths of light.

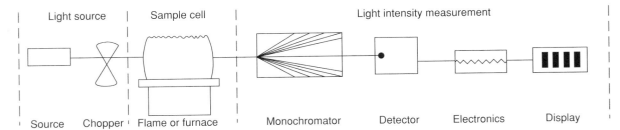

Figure 9 A schematic of an atomic absorption instrument

The way in which atomisation takes place, and therefore the design of the atom cell, accounts for the difference between types of atomic absorption spectrometers.

The light source
The light source in atomic absorption spectrometers emits specific wavelengths that are absorbed by the atoms of the analyte. An excited metal is the source and emits the same spectrum of wavelengths as that absorbed by the analyte (*ie* the source and analyte are the same element). This phenomenon is known as the 'lock and key' effect. If an analysis for the concentration of lead atoms in water is being done, the water sample contains the analyte, and a lamp consisting of excited lead atoms is the emitting source.

Atomic absorption spectrometry (AAS) is selective because no two elements absorb radiation at exactly the same wavelengths. Even where two or more elements may absorb strongly at wavelengths very close to each other – *ie* within 1 or 2 nm, other absorption wavelengths will help to establish a 'fingerprint' that can distinguish between them.

RS•C

The hollow cathode lamp (HCL)

The most widely-used source of light is a hollow cathode lamp, (HCL). This lamp contains a tungsten anode and a cylindrical cathode made of the element to be analysed. The anode and cathode are sealed in a glass tube filled with an inert gas (either neon or argon), called the filler gas, at a low pressure (ca 1.5 kN m^{-2}). When a potential difference of ca 300 V is applied between the cathode and the anode, some gas atoms ionise. These positive gas ions accelerate towards the cathode, and on impact these ions eject metal atoms from the cathode – a process called 'sputtering'. Through bombardment with ions of the filler gas some of the sputtered atoms become electronically excited, and emit radiation characteristic of the metal as they fall back to the ground state.

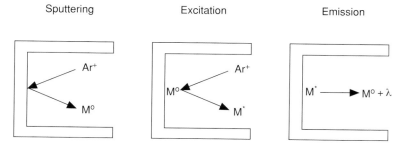

Figure 10 A typical process in the HCL. Ar$^+$ is a positively-charged argon ion, M^o is a sputtered ground-state metal atom, M* is a metal atom in an excited state, and λ is radiation emitted at a wavelength characteristic of the sputtered metal

This radiation needs to pass through to the detector, hence the position of the lamp and the shape of the cathode are crucial. The concave shape of the cathode concentrates the beam which passes through a quartz window designed to allow through as much light as possible. Light emitted from the lamp passes through the atomised sample and then to the detector. Most of the atoms are redeposited on the cathode.

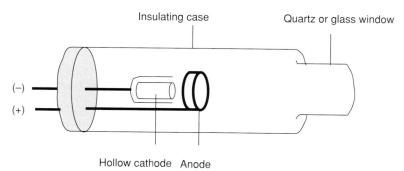

Figure 11 The structure of a hollow cathode lamp

Modern absorption instruments have several lamps, each made from a different element, housed in a rotating turret so that the correct lamp can be selected for the element that is being analysed.

RS•C

The filler gas **Box 3**

The filler gas has to be chemically inert, with a high ionisation energy. The gas should absorb at wavelengths which have minimal interference with the spectra emitted from most metal sources. Neon and argon are the most common filler gases.

The choice between neon or argon depends on the metal used in the cathode. Neon is used because it has a simpler spectrum than argon, hence argon is only used with metals where there is spectral interference from neon.

Problems with hollow cathode lamps

After sputtering, some metal atoms tend to diffuse away from the cathode without being in an excited state. As a result these atoms absorb radiation from the lamp before the light reaches the atom cell.

Another problem is the deposition of the metal on the window of the lamp after the instrument has been used for some time. This leads to a reduction in the intensity of light from the lamp. This can be compensated for by increasing the separation between the cathode and the window. An alternative to the HCL is the electrodeless discharge lamp that uses the energy of a radiofrequency field to vaporise and excite the atoms of the metal in the bulb. Electrodeless discharge lamps are occasionally used for the more volatile elements such as arsenic, bismuth and antimony.

Modulating the light beam

The detector has to distinguish between light from the HCL and from the sample. A rotating disc called a 'chopper' modulates – *ie* rapidly switches on and off – the radiation from the lamp. The detector is programmed to select modulated light as that emitted from the lamp.

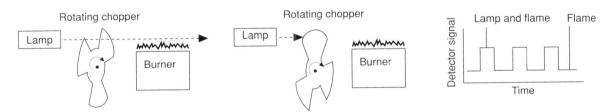

Figure 12 The effect of a rotating chopper on the detector signal

RS•C

Double beam spectrometers

Modern atomic absorption spectrometers incorporate a 'beam splitter' so that one part of the beam passes through the atom cell and the other is a reference.

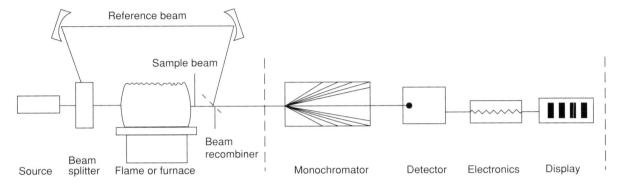

Figure 13 A schematic of a double-beam atomic absorption spectrometer

The intensity of the light source cannot always be relied upon to remain constant during an analysis. If only a single beam is used to pass through the atom cell, a 'blank' reading containing no analyte would have to be taken first, setting the absorbance at zero. If the intensity of the source changes slightly by the time the sample was put in place, the measurement is inaccurate. In the double beam system there is constant monitoring between the reference beam – effectively acting as a blank – and the light source. So that the spectrum does not suffer from loss of sensitivity, the beam splitter is designed so that as high a proportion as possible of the energy of the lamp beam passes through the sample.

The atom cell

The main difference between atomic absorption spectrometers is the way in which atoms are generated in the atom cell. This difference is reflected in the two common forms of AAS. In flame AAS (FAAS) a solution of the sample is aspirated into a flame where atomisation and absorption take place. Graphite furnace AAS (GFAAS) involves heating the sample in a graphite tube, producing a 'cloud' of atoms. Absorption occurs as the beam of light is shone along the axis of the tube.

Flame atomic absorption spectrometry (FAAS)

Flame atomic absorption spectrometry is the most routine and widely used form of AAS. The sample solution is transported along a flexible capillary tube before being nebulised – *ie* broken into tiny droplets – in the burner. Larger drops settle out and drain off while smaller ones vaporise in the flame. Only about 1% of the sample is atomised, hence FAAS is unsuitable when only a small amount of sample is available for analysis.

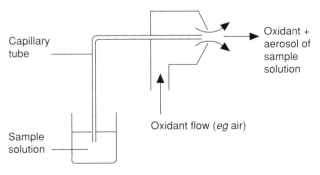

Figure 14(a) The principle of a nebuliser

RS•C

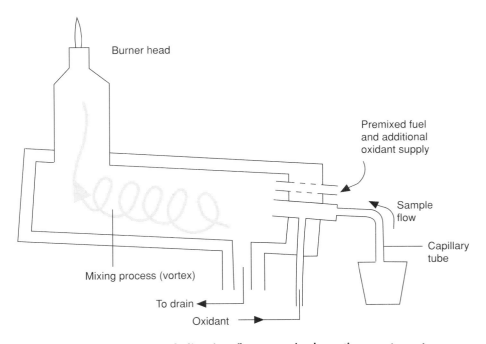

Figure 14(b) The nebuliser in a flame atomic absorption spectrometer

The nebuliser

The nebuliser consists of a capillary tube that runs between the sample solution and the burner. An automated sampler rotates, presenting each sample or standard for analysis in turn. The nebuliser head is made of a corrosion resistant material such as a platinum alloy because, sometimes, the sample is dissolved in acids that would quickly corrode other materials.

Nebulisation creates a fine mist or aerosol of the sample solution in the burner where it mixes with the fuel and oxidant gases. The heat of the flame vaporises the solution to dryness and dissociates molecules into free atoms in the ground state.

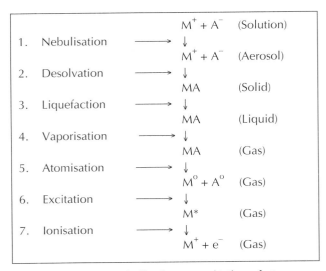

Table 1 From nebulisation to excitation of atoms
(Ionisation tends to occur with metals of a low first ionisation energy
– *eg* the alkali metals)

RS•C

Efficient nebulisation is crucial because the droplets have to be small enough to be vaporised and atomised during the short time they spend in the flame. Larger droplets cool the flame and hinder atomisation because too much of the solvent has to be vaporised.

The time spent by the sample in the flame is about 0.025 sec so it needs to be atomised as quickly as possible. The oxidant gas – usually air or nitrogen(I) oxide (N_2O) – flows quickly across the end of the capillary tube. This results in the pressure dropping below atmospheric pressure, and the liquid rises up the capillary. The following sequence of events takes place.

The solution emerging at the tip of the capillary forms a mist because of the turbulence caused by the fast flow of the oxidant gas. The formation of an aerosol mist is further assisted as the solution emerging from the tip of the capillary is directed at high speed towards a glass bead.

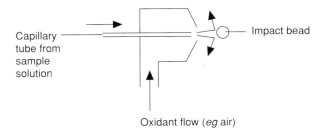

Figure 15 Impact bead and aerosol spray

The aerosol is mixed with the fuel and oxidant gases before flowing into the burner.

Baffles enable further mixing and block any large droplets of liquid, which are filtered out into a trap and then drained off.

The burner
Air-ethyne (air is the oxidant gas and ethyne is the fuel) flames operate at temperatures between 2100 °C and 2400 °C and are used most widely in FAAS. They are hot enough to provide ground state atoms of most analytes. A slot burner is used to maximise path length through the flame carrying the analyte.

Beer's law shows that maximising path length (l) increases sensitivity, particularly important at low concentrations.

($A \propto cl$; so increasing path length (l) means an increase in absorbance (A))

In practice, slot burners have a width of 1 cm. They are usually made of titanium, which resists heat and corrosion.

The analyte is drawn up into the flame where the atoms absorb light from the lamp. Where the air-ethyne flame is not hot enough to atomise a sample, the nitrogen(I) oxide-ethyne flame is used. This is particularly useful for the refractories – oxides of metals such as antimony, aluminium and molybdenum. Refractory materials can withstand extremely high temperatures without chemical change. The average temperature of a nitrogen(I) oxide – ethyne flame is about 3000 °C.

RS•C

Safety   **Box 5a**

Ethyne is explosive and flammable, and appropriate safety precautions have to be taken. In laboratories operating a FAAS there has to be a gas detection system for hydrocarbons.

The nitrogen(I) oxide-ethyne mixture produces a richly burning flame, but there is risk of carbonisation occurring in the burner. Carbon particles can fall back into the burner leading to explosions. This risk can be minimised by lowering the flow rate of gases through the burner.

It is important to ensure that the nebuliser provides a steady flow of analyte to the flame to prevent over-heating.

Methods of sample preparation   **Box 5b**

Samples that are analysed using atomic spectrometry have a variety of different origins including air, water, rock, food, beverages, plant material, detergents and clinical materials. Some of these sources, such as water and beverages need little preparation. The main preparations involve neutralisation of some acidic drinks, or concentrating the sample where the analyte is present in a very dilute form. Collecting air-borne samples poses greater problems but dust particulates can be deposited on filters or electrodeposited on columns. Some pre-concentration may be necessary such as:

■   complexing certain selected metals and extracting them from solution; or

■   preferentially collecting samples on columns followed by elution with acid.

Preconcentration is often used for clinical samples.

If the concentration of the analyte is too high for the calibration range (see below) then the following measures can be taken:

■   diluting the sample;

■   rotating the burner to reduce the path length (l); or

■   using alternative wavelengths to reduce sensitivity.

Acid digestion is most commonly used for solid samples usually with the assistance of a microwave oven. However, this is not always straightforward. For example, iron dissolved in nitric acid is precipitated by silicates or phosphates in plant material. Higher yields of iron are obtained if a mixture of nitric and chloric (VII) (perchloric) acids are used for digestion rather than nitric acid alone. Chloric (VII) acid can cause explosions when solutions are evaporated to dryness, so appropriate precautions must be taken.

Wet digestion is accelerated using a microwave oven. The sample is placed in an air-tight container and subjected to high frequency microwaves at an increased pressure and temperature of 100–300 °C. While microwave digestion is very quick this is offset by the time taken for the vessel to cool down before it can be handled.

Geological samples contain species that are very difficult to dissolve. Rocks and silicates can be fused in platinum crucibles with an appropriate flux, usually lithium metaborate ($LiBO_2$). The melt is leached with dilute acid. However, using digestion or fusion means that more volatile components may be lost.

Sample preparation

Some samples, such as solutions in water and biological fluids, can be aspirated directly into the flame, others such as solids must be dissolved in acid. Methods of sample preparation, and the uses of acids for analytical purposes, are discussed more fully in *Sampling and sample preparation*.

RS•C

Reagent	Sample types	Examples	Special features
Nitric acid	Most samples	Analysis of metals in detergents	Virtually all nitrates are soluble at the concentrations used for analysis.
Hydrochloric acid	Most samples	Analysing alloys and some metal oxides	General use acid preferred to nitric acid as a solvent because it is less hazardous in the concentrated form.
Hydrofluoric acid	Silicates	Analysing rocks	It attacks glass and any silicates. Plastic vessels should be used.
Ashing	Plant material, coal, resins, biological material, municipal waste, meat	Analysis of metals concentrated in leaves	Dry ashing involves heating the sample in a furnace for up to 24 hrs, then extracting the residue with acid. Wet ashing is quicker and involves using oxidising acids such as concentrated nitric acid.
Fusion	Intractable materials such as rocks that do not dissolve in acids	Clays	The sample is mixed with lithium metaborate ($LiBO_2$) at 1100°C. A solid solution is formed. This can then be dissolved in concentrated mineral acids.

Table 2 Methods of sample preparation for FAAS

Case Study

Determining magnesium concentrations in mineral waters

The purpose of this analysis is to monitor the concentrations of magnesium that appear on the labels of two bottles of mineral water. These are:

A   Mg   24 mg l^{-1} (2.4×10^{-2} g dm^{-3})

B   Mg   6.1 mg l^{-1} (6.1×10^{-3} g dm^{-3})

The stock solution for preparing magnesium standards for the calibration curve contains magnesium at a concentration of 1×10^{-1} g dm^{-3}. Suitable dilutions are made of the stock solution – ie 0.2, 0.5, 1, 1.5, 2 mg dm^{-3} – by making up 100 cm^3 volumes using volumetric flasks.

A magnesium HCL is used as the light source, and a setting of 285.2 nm is made for the absorption wavelength of magnesium. Deionised water is aspirated and the flame is lit. (An analyte solution or deionised water must always be aspirated while the flame is on.) An air-ethyne flame is used and adjusted until a reading of zero absorbance is obtained. A standard solution (0.5 mg dm^{-3}) is used and the flame is adjusted until highest absorbance is achieved, which remains steady for ca 5 secs. Deionised water is sprayed into the flame between measurements and occasionally checked for zero absorbance. The results are tabulated in the *Questions* section (p. 24).

Interference

Samples contain the analyte plus the matrix. The analyte is that part of the sample that is measured, and the matrix is the remainder of the sample – ie it consists of all the other components. For example, when determing the sodium chloride content of rock salt the sodium chloride is the analyte and any other components form part of the

RS•C

matrix. Interference arises from any species in the matrix that affect measurement of the analyte.

Ionisation suppression
Metals that have low first ionisation energies, such as the alkali metals (sodium, potassium etc) tend to ionise in the flame. The absorption wavelengths for sodium ions differ from sodium atoms, but problems can be overcome by adding a large amount of an element with an even lower ionisation energy to the sample solution. In this process the ionisation is considered as an equilibrium system:

$$Na(g) \rightleftharpoons Na^+(g) + e^-$$

$$K(g) \rightleftharpoons K^+(g) + e^-$$

Potassium ionises more easily than sodium. The greater availability of free electrons from the ionisation of potassium shifts the sodium ionisation equilibrium to the left and suppresses the formation of sodium ions.

Preventing association
Metals such as calcium may form stable compounds, even when aspirated into the flame, particularly with polyatomic anions such as sulfates and phosphates, *eg*

$$3Ca^{2+}(g) + 2PO_4^{3-}(g) \rightarrow Ca_3(PO_4)_2(g)$$

Finer aerosols and hotter flames can prevent association. Releasing agents can be used to form complexes with the anion, leaving the metal free – *eg* by adding metals such as lanthanum as releasing agents which form thermally stable compounds with the anion more readily than calcium.

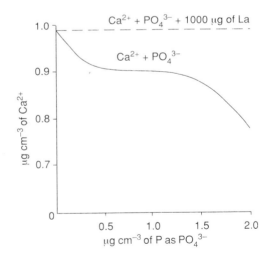

Figure 16 The effect on calcium absorbance of adding PO_4^{3-}
The dotted line shows the effect of adding La as a releasing agent.

Calibration

A calibration curve is drawn using a range of known concentrations of the analyte, called standards, and determining the absorbance for each concentration. For accurate calibration:

■ a blank solution is prepared for zero absorbance. Any matrix modifiers such as ionisation suppressors are added to the blank as well as to all samples;

RS•C

- the samples should be fresh and prepared from a concentrated solution. The preparation flask is acid-washed to prevent adsorption of metal ions onto the glass surface;

- reagents should be pure, particularly for trace element analysis. Preparation flasks, such as volumetric flasks, should be soaked in the acid solvent first. It is best to use polytetrafluoroethene (PTFE) coated surfaces because they adsorb fewer metal ions than glass, and acid can leach metals from glass;

- only the linear parts of the calibration plot should be used;

- the range of concentrations of standards should exceed the concentration range of the samples. Five standards at a wide range of concentrations should be sufficient for a calibration curve; and

- absorbances should be in the range of 0.1–1.0, and the most suitable wavelength should be chosen to give this range.

Standard additions

The response of FAAS to the concentration of the analyte may not be determined by the analyte alone but by matrix interference. For example, if a sample dissolved in a viscous acid such as phosphoric acid has a lower nebulisation efficiency than an aqueous solution, less analyte per unit time will enter the flame, compared with an aqueous solution of the standard.

Standard additions correct for some matrix interference by adding a standard of the analyte to the sample. In this method the standard is affected by the same interference as the sample. In standard additions, three equal aliquots of the sample are taken.

- Nothing is added to the first aliquot.

- An equal volume of standard is added to a second aliquot of the sample. (The concentration of the analyte in the standard is approximately equal to the estimated concentration of the analyte in the sample.)

- Twice the volume of standard is added to the third aliquot.

All three portions are then made-up to the same volume so the concentrations of all the components in the original sample are the same in each case. The absorbances of the three portions are recorded and compared with an aqueous calibration plot.

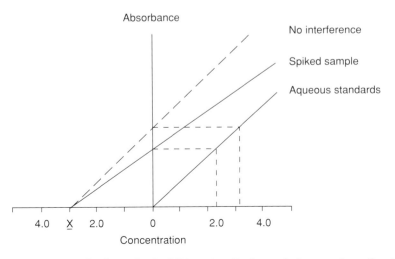

Figure 17 A graph of standard addition. A spiked sample is one where the standard has been added to the sample. X indicates the concentration of the analyte.

If there is no interference in the sample, the slope of the standard addition line is the same as that for an aqueous standard. There is interference if the slope of the standard addition line is different from that of the aqueous calibration plot. The concentration of the analyte is calculated by continuing the line to intersect the concentration axis.

One important limitation of the standard addition method is that it does not compensate for chemical and ionisation interference.

Limitations of FAAS

Although FAAS continues to be the form of AAS that is most widely used, there are limitations to the technique. These include:

■ the relatively short time that the sample spends in the flame;

■ unsuitability for small samples. Only about 1% of the sample is actually atomised and much of the analyte is lost as large drops;

■ the need for solids to be dissolved or extracted, which can result in losing the sample;

■ the absorption by molecular species, such as ethyne and oxygen, in the flame at around 200 nm. This is a problem for elements, such as arsenic, whose most prominent absorption wavelength is in this region; and

■ the range of concentrations of an analyte (the linear dynamic range, LDR) that FAAS can measure is small compared to other atomic spectrometric techniques.

Graphite furnace atomic absorption spectrometry (GFAAS)

The main alternative to FAAS is Graphite Furnace Atomic Absorption Spectrometry, GFAAS. The main advantages of GFAAS are:

■ the atoms are present in the atom cell for a relatively long time;

■ there is no flame so it is safer than FAAS;

■ no nebuliser is needed;

■ there is a 100-1000 fold decrease in the detection limit over FAAS;

■ there is little or no loss of material during atomisation, so small quantities of sample can be used. This is particularly useful for samples such as blood where only a small amount can be removed from a patient; and

■ it can be used to analyse solids.

Element	FAAS Detection limit ng cm^{-3} (ppb)	GFAAS detection limit ng cm^{-3} (ppb)
Arsenic	100	1.0
Barium	8	0.2
Cadmium	1	0.01
Lithium	0.5	1.0
Molybdenum	10	0.06
Selenium	100	0.2
Zinc	0.6	0.005

Table 3 Detection limits of FAAS and GFAAS for a range of elements

RS•C

Design for GFAAS

In GFAAS the sample is placed in a graphite tube which is heated electrically. The graphite material resists the passage of an electric current of about 100 amperes and converts the electrical energy into heat.

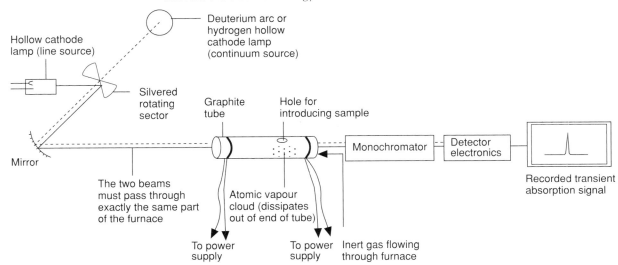

Figure 18 A schematic of a GFAAS instrument showing the furnace in relation to the optics

The graphite tube is typically about 30 mm long and 10 mm in diameter. The sample is usually injected using a micropipette through a hole at the top of the tube and is atomised a few seconds after the tube is heated. A beam of light of the required wavelength is directed along the axis of the tube. After analysis the sample is flushed from the tube by argon.

Adjusting for GFAAS

Background absorption (Box 6)

While GFAAS overcomes the problem of interference in FAAS of molecular absorption in the flame, the greater time spent by the sample in the tube increases the chances of absorption from other sources – eg particulate matter from the matrix. (There tends to be more background absorption in GFAAS, particularly with biological and geological samples due to residual organic material and vaporised matrix salts.) Particle scattering – eg the vaporisation of halide salts as molecules, prevents light from reaching the detector. Halide salt absorption increases sharply below 220 nm in the deep ultraviolet region. This is a problem for elements that absorb below 220 nm – eg lead, cadmium and arsenic.

The problem of background absorption can be addressed by furnace programming – ie preparing the furnace to eliminate as much background material as possible before atomisation.

There are three main sequences in furnace programming.

- Drying-off the analyte solvent in the graphite tube at a temperature of ca 120 °C. Excess solvent is flushed out using argon.

- Ashing at 600–800 °C to break down organic matter into smaller molecules which can be flushed more easily from the furnace.

- Atomising at 2000–2500 °C to produce a rapid absorption peak.

RS•C

The gas in the tube is cooler than the heated walls, hence another problem is condensation. Atoms forming on the wall of the tube migrate rapidly into a cooler gas and associate to form molecules or condense back onto to the surface of the walls. Most graphite furnace tubes have a L'vov platform, which consists of a raised graphite platform on which the sample is placed. This ensures that the air in the tube and the sample are at the same temperature. The sample is then heated indirectly by radiation and convection from the furnace walls. This gives the surrounding gas a chance to heat up and it is only at this point that the atoms migrate into the warmer gas, so preventing condensation.

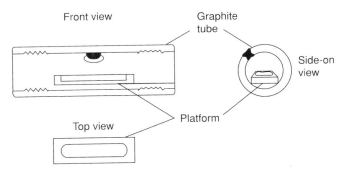

Figure 19 A L'vov platform

Matrix modifiers

Problems with both the matrix and the analyte can arise in GFAAS, particularly during furnace programming. Matrix modifiers are added to diminish any interference. These are reagents that are added to the sample to modify the behaviour of the matrix or, occasionally, the analyte. Examples where modifiers can be used are:

■ stabilising the analyte during the ashing stage to permit a higher ashing temperature (volatile analytes such as arsenic can escape from the tube during ashing); and

■ converting an interfering matrix into a volatile compound which can then be removed during ashing.

A small quantity of palladium (ca 0.1% m/v) is used to stabilise volatile elements such as antimony, arsenic, bismuth and selenium. In the absence of palladium the ashing temperature has to be kept below 500 °C to avoid losing these elements as analytes. Palladium forms stable compounds with these analytes, allowing a substantial increase in the ashing temperature to drive-off any interfering organic matrices.

Hydride generation atomic absorption spectrometry (HGAAS)

Some elements, predominantly in groups five and six of the Periodic Table, such as arsenic, antimony and selenium, form gaseous covalent hydrides. This property can be utilised because the hydrides can be separated from the sample matrix before being introduced into the light path. The hydride is decomposed and atomised with negligible background interference. This technique has considerable advantages for analysing volatile elements such as arsenic. In the food industry, arsenic has to be detected at levels that are below the detection limits for FAAS. On the other hand, arsenic is too volatile for GFAAS and escapes from the atom cell before a measurement can be made. The detection limits for HGAAS are lower than those for FAAS.

RS•C

Background correction for GFAAS and FAAS   **Box 6**

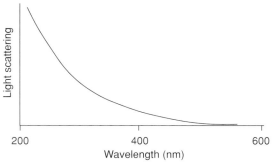

Figure 20 The wavelength dependence of light scattering

A clear and unambiguous read-out is obtained only when the analyte atoms absorb the incident light. In practice, there is always some background interference, particularly in furnace techniques, from smoke particles and from impurities on the wall of the tube. Undissociated molecules in the matrix may have broad band spectra overlapping that of the analyte. Some background interference is introduced with the sample itself and this cannot be corrected by comparing with the blank. The degree of light scattering is inversely related to the wavelength of the source – shorter wavelengths are scattered more than higher wavelengths. In practice, a background correction for interference is necessary for wavelengths below 350 nm.

There are two main systems used in AAS to correct for background absorption. These are the continuum source technique and the Zeeman technique. They work on a similar principle. One measurement is made of the analyte and background to give the total absorbance. A second measurement is made where analyte absorbance is reduced to a negligible amount so that only the background is measured. Subtracting the second measurement from the first gives the true analyte absorbance.

The continuum source technique

Two lamps are used in this system to correct for background scatter – the HCL that emits a narrow band of wavelengths and a deuterium arc lamp. An electrical discharge in deuterium gas emits light over a much wider range of wavelengths than that of the HCL. To help distinguish between them a chopper passes the beams alternately through the atom cell.

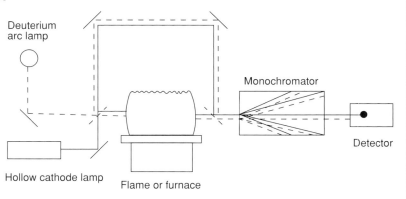

Figure 21 A continuum source arrangement

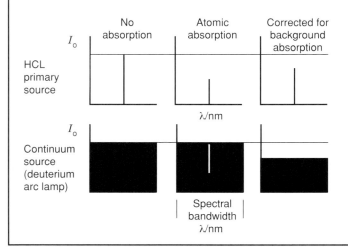

The intensity of light from the continuum source is considerably reduced in the atom cell by the overall background, but negligibly reduced by the sharp absorption wavelength.

Figure 22 (top) Atomic and background absorption with a primary (line) source

Figure 23 (bottom) Atomic and background absorption with a continuum (broadband) source

Continued...

RS•C

Box 6 continued: Background correction for GFAAS and FAAS

An electronic signal is generated in the detector, which compares the two beams. A response is displayed only where the absorbance of the two lamps differs.

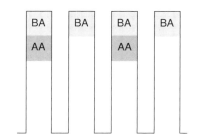

Figure 24 Square wave signals showing a continuum measuring background absorbance (BA) and a primary source measuring both background absorbance (BA) and analyte absorbance (AA)

Limitations of the continuum source technique are:

■ two light sources are required which need occasional replacement; and

■ the two sources have to be accurately aligned so they view exactly the same portion of the atom cell. This is difficult to monitor.

There is a continuous sample throughput in FAAS, maintaining the same conditions over a period of time so background correction with this system is accurate and easy to implement. However, in GFAAS the background absorption changes over time because the atom cell undergoes sequential changes when the atom cloud is formed.

The hydride is prepared by adding the sample to a solution of hydrochloric acid (5 mol dm^{-3}), then reducing with sodium tetrahydridoborate(III) (NaBH$_4$). The hydride vapour is then flushed into the silica tube using argon gas. The tube is pre-heated to 900 °C which is sufficient to atomise the vapour. In modern instruments the metal hydride is decomposed in an electrically-heated quartz cell.

Cold vapour atomic absorption spectrometry (CVAAS)

Cold vapour atomic absorption spectrometry (CVAAS) is a technique used to determine only mercury. The main absorption wavelength for mercury is at 184.9 nm. When using flame methods, less sensitive lines have to be used because of the proximity of this wavelength to the maximum absorption wavelengths of flame gases. Mercury has a vapour pressure at room temperature that is high enough to make a sample suitable for analysis without heating, hence a specialised technique has been devised for determining mercury in a sample.

Any mercury compounds in the sample can be reduced using sodium tetrahydridoborate(III) or tin(II) chloride (SnCl$_2$).

$$Hg^{2+}_{(aq)} + Sn^{2+}_{(aq)} \rightarrow Hg_{(l)} + Sn^{4+}_{(aq)}$$

Organic mercury compounds are oxidised to inorganic mercury before reduction.

Argon gas is bubbled through the solution to flush out the mercury that is put into a silica tube. Light is then transmitted along the axis of the tube from a hollow cathode mercury lamp, and measurements are made.

RS•C

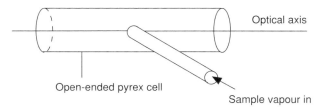

Figure 25 A mercury vapour cell

Comparison	FAAS-air-ethyne flame	FAAS-NO-ethyne flame	GFAAS	CVAAS (Hg only)
Cost	Low	Medium	Very high	Medium
Elemental scope	Good	Very good	Good	Poor
Speed	Good	Good	Medium	Medium
Sensitivity	Medium (ppm)	Medium (ppm)	Very good (ppt)	Good (ppm or ppb)

Table 4 Comparing AAS techniques

The detecting components

Monochromator

A monochromator is a device that isolates the measured wavelength from other emission wavelengths. Conventional monochromators incorporate:

- an entrance slit for the light;

- a diffraction grating to disperse the different wavelengths; and

- an exit slit.

The optics are adjusted so that light of the selected wavelength emerges through the exit slit of the monochromator towards the detector.

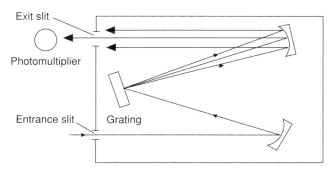

Figure 26 A monochromator

Dispersion – *ie* the extent to which the wavelengths are separated – is related to the number of lines on the diffraction grating. Greater dispersion is achieved by using a larger number of lines per unit width of the grating. With a low dispersion grating the exit slit has to be narrow to ensure that the detector receives light of the selected wavelength. A high dispersion grating means that the exit slit can be wider, so allowing a greater proportion of light to reach the detector, thereby increasing sensitivity.

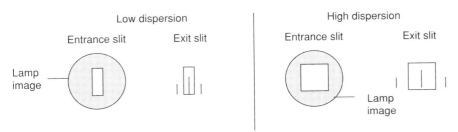

Figure 27 The influence of dispersion on the width of the exit slit

Another important design factor is the blaze angle, the angle etched on the surface of the grating. The grating disperses the wavelengths, but the wavelength of light that loses least in intensity is where the angle of reflection equals the angle of incidence.

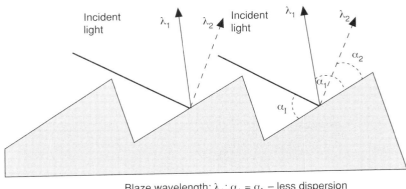

Blaze wavelength: λ_1: $\alpha_1 = \alpha_I$ – less dispersion
λ_2: $\alpha_2 \neq \alpha_I$ – more dispersion

Figure 28 Grating blaze angle

A grating can be machined to control the blaze angle. Since the range of absorption wavelengths is *ca* 200–900 nm, a blazing cut for wavelengths in the middle region means significant reductions in intensity for wavelengths near 200 nm or 900 nm. Most spectrometers incorporate two gratings, one blazed in the ultra-violet region and the other blazed in the infrared region.

Detectors

Detectors measure the intensity of the reference beam and the intensity of the light that emerges from the sample. The most common form of detector is a photomultiplier. This consists of a sealed tube covered by a photoemissive surface – ie it converts light in the form of photons to an electrical signal.

This electrical signal is amplified by successively accelerating the electrons through a series of anodes, up to a potential of 100 V. The ejected electrons are accelerated towards the first anode which then ejects two to five times as many electrons as those incident on it. Given that multiplier tubes contain up to 16 of these stages, as many as a million electrons are generated by a single photon. The resulting electric current is related to the intensity of the light radiation reaching the detector.

RS•C

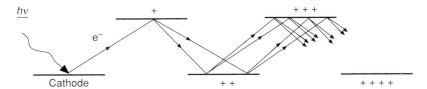

Figure 29 Successive amplification of electrons

The surface should contain a material that is easily ionised. For most purposes, a caesium-antimony alloy is used for metals with absorption wavelengths in the ultraviolet and visible part of the electromagnetic spectrum, but not for the infrared. For longer wavelength studies, up to 900 nm, trialkali-antimony surfaces and gallium arsenide (GaAs) are used.

Signal processors

Signal processing is the conversion of the electrical signal generated by the photomultiplier into information that can be used by analysts. Photomultiplication produces direct current signals which are susceptible to poor sensitivity and low signal to noise ratio. An alternating current (AC) signal is produced by modulating the light beam.

By using a phase-sensitive detector responding only to signals of the same phase, all noise at other frequencies can be eliminated. Information is then processed through a computer into a form that analysts can use.

Applications

Atomic absorption spectrometry is a quick and sensitive method for determining the concentrations of up to five elements at a time, usually metals. It is widely used to monitor contamination in agricultural products, foods, detergents and pharmaceuticals. For example, copper appears as a trace element in ammonium nitrate fertiliser and is a catalyst for a series of reactions that causes ignition. Samples of the fertiliser are determined for copper using FAAS. A higher limit is set at 10 ppm, although it takes much higher concentrations to promote ignition.

Copper concentrations are monitored in a variety of products using AAS. Copper ions catalyse the oxidation of oils. This is a problem in the food industry where 'off-flavours' may occur in products such as margarine and liquid vegetable oils. Trace amounts of copper – as low as 0.005 ppb – can cause oxidation. This requires a very sensitive detection method, hence GFAAS is used rather than FAAS.

Companies also use AAS to investigate competitors' products, either for possible infringements of patents or to make guesses about the possible make-up of new products.

Atomic absorption spectrometry can be used in problem-solving, again in the food industry. For example, a new brand of tea may be causing unexpected precipitates. Atomic absorption spectroscopy is used to identify what is precipitating out, then the raw materials are investigated to trace the origin of the contaminant.

Zinc citrate prevents tooth decay. Flame atomic absorption spectrometry is used to monitor zinc levels in toothpastes so they remain within certain pre-defined concentration limits. (Adding too much zinc citrate raises costs whereas too little has no effect.)

Soap tends to go rancid when the concentration of iron or copper exceeds a certain level and AAS is used to monitor the levels of these metals.

RS•C

There is a wide variety of other applications of AAS including identifying trace metals in babies' dummies, nickel in the air where welding is done, biomonitoring pollen for heavy metal pollution and investigating senility through the distribution of trace metals in the brain with increasing age. Cold vapour techniques are used to determine mercury levels in workplaces such as chlor-alkali plants and in saliva when dental fillings corrode.

Atomic absorption spectrometry is also used in many laboratories to analyse samples for toxic metals from potentially contaminated soils – *eg* old gas work sites. Samples from agricultural land on which sewage sludge is spread is also analysed to ensure the land is not contaminated with heavy metals.

Atomic absorption spectrometry is used widely in standardising for legislation. For example, nickel is added to some jewellery. It is released on to the skin on prolonged contact and can cause an eczema-like rash. Under new European Union legislation on products that are worn on the skin, the rate of release of nickel in one week must not exceed 0.05 μg cm^{-2} in products such as earrings, bracelets, watches and necklaces.

The root of the problem

Hair is the metabolic end product of lead in the body, and it has been used to measure lead levels in children living near motorways, for example. It provides a way of estimating lead intake without the complex methods of sampling body fluids such as blood or urine. One chemist decided it would be a good idea to test his family.

His immediate family and his brother's family lived in modern houses with copper piping in the plumbing system. His mother and father lived in an older property in the same area which had lead piping. Was any of the lead being leached from the pipe work into their water supply and into his parent's bodies?

He took hair samples from the members of each family, dissolved the samples in concentrated nitric acid to convert the lead to lead nitrate, and measured the lead concentrations by atomic absorption. The lead levels in his parent's hair were significantly higher than those in his brother's hair, as expected. However, one unexpected result was that his father's hair had lead levels of over twice those recorded for the mother (230 ppm compared with 92 ppm). The explanation for this odd result was later found to be that his father, without the knowledge of the rest of his family, was using the hair dye Grecian 2000 which contains lead(II) ethanoate.

RS•C

Questions

Flame atomic absorption spectrometry was used to check the magnesium concentration in two brands of bottled water, (A and B). The concentrations listed on the labels are:

A) Mg 24 mg dm^{-3}

B) Mg 6.1 mg dm^{-3}.

The instrument was calibrated using standards. The following results were obtained:

Concentration of Mg standard /mg dm^{-3}	Absorbance
0.2	0.062
0.5	0.254
1	0.417
1.5	0.690
2	0.801

Two bottled water samples were diluted and then analysed. Sample A was diluted 20 times and sample B five times. The absorbance readings are:

A) 0.635

B) 0.624

1. a) Draw the calibration graph.

   b) How satisfactory are the results in terms of quality control? What factors may have affected the accuracy of the results?

2. A detergent manufacturer monitors arsenic levels in toothpaste. Silica is also present in toothpaste. Suggest a method, using AAS, for monitoring the arsenic.

Answers

1a) Calibration graph.

 b) The concentrations of the two samples are greater than those of the standards. Dilutions had to be carried out. Factors such as flow rate and temperature may have affected the calibration and the dilution will have lowered sensitivity.

2   The best method for determining arsenic is by HGAAS. Silica is a problem as a matrix because the best way of breaking it down is by fusion and dissolution. However, arsenic is volatile and is lost in the fusion process.

   The problem becomes more tractable if placed in the context of the way in which toothpaste is used. Above a certain level arsenic is toxic in the body. Silica is an abrasive and is not absorbed, so there is no need to matrix match and determine the arsenic concentration in the presence of silica. The arsenic can be extracted using 10% v/v hydrochloric acid solution.

RS•C

Inductively coupled plasma (ICP) techniques

Inductively coupled plasma – atomic emission spectrometry (ICP–AES)

The technique of linking atomic emission spectrometry (AES) (Figure 1) to an inductively coupled plasma (ICP) is given the acronym ICP-AES. In contrast to atomic absorption spectrometry (AAS), ICP–AES can identify and determine the concentrations of up to 40 elements *simultaneously* with detection limits of parts per billion. (1 part per billion (ppb) is 1 μg dm^{-3} or 10^{-6} g dm^{-3}). The ICP torch is a device that produces a plasma – a fireball of atoms, ions and electrons interacting at very high energies with temperatures up to 10 000 K. It is a very effective atomisation source! Exposing a sample to a high temperature plasma converts a very large proportion of its constituent atoms to an excited state. The atoms emit electromagnetic radiation as electrons return to a lower energy state. This means that ICP–AES has one important advantage over atomic absorption: the sample itself acts as a light source, emitting electromagnetic radiation. When a sample consists of more than one metal there will be multiple light sources. By using suitable polychromators and scanning monochromators, multi-element analysis is possible (this is explained later in the chapter).

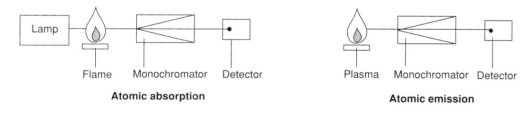

Figure 1 The comparison of atomic absorption spectrometry (AAS) and atomic emission spectrometry (AES)

In atomic absorption, electrons are promoted only from the ground state. However, an emission spectrum is often more complex than an absorption spectrum because emission can occur to other excited states (Figure 2).

RS•C

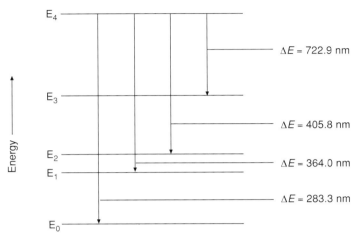

Figure 2 Some emission transitions for lead

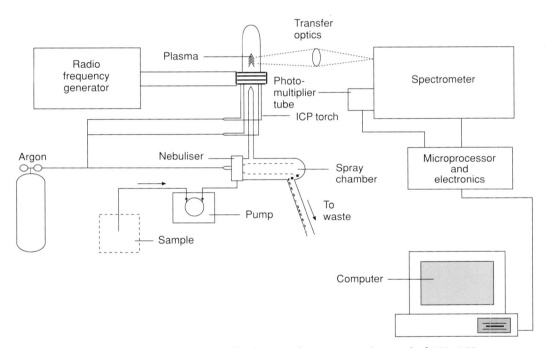

Figure 3 The layout of components in a typical ICP–AES

In modern instruments, after dissolution of the sample, the ICP–AES analysis is fully automated whereupon the sample solution is usually transported to the plasma. Once nebulised and inside the plasma the sample is vaporised, atomised and ionised. The radiation resulting from emission is transferred directly to a spectrometer where the various wavelengths are sorted optically, electronically detected and analysed. Sample throughput normally involves four stages:

■ the time for the sample concentration to reach the plasma and for the signal to stabilise;

■ taking readings;

■ a background correction; and

■ a print out of results.

A normal analysis takes 2–6 mins per sample.

RS•C

The sample aerosol is introduced into the centre of the plasma forming a stable toroidal (bun-shaped) plasma. The high temperatures produced by the plasma and the long residence time (*ca* 2 milliseconds) of the analyte in the plasma mean that the analyte is completely atomised. All the atoms are in an excited state, hence no self-absorption occurs – *ie* there is maximum sensitivity, because there are no atoms in the ground state to absorb the emitted electromagnetic radiation.

Box 1

The production of atoms in refractory oxides – oxides of metals such as antimony, aluminium and molybdenum – is a problem in atomic absorption spectrometry (AAS) but the high temperature of the plasma ensures that all oxides are dissociated. Chemical interference is overcome because all molecular species are atomised at high temperatures. The large number of emission lines for each analyte means that an appropriate line can be selected, giving interference-free analyses for most analytes. For example, although titanium interferes with cobalt for one emission line, another can be chosen for cobalt.

Calibration

Calibration is performed using low-pressure mercury discharge lamps with well-defined emission lines in the ultraviolet/visible region. The calibration for concentration versus signal strength is done by using solutions of known concentration, and the lack of self-absorption facilitates the production of linear calibration curves. Calibration is linear over four to six orders of magnitude (1×10^4 or 10^6) so typically only two standards and a blank are needed because it is a linear calibration over a wide range. The calibration curve is stored in a computer and corrections are made automatically.

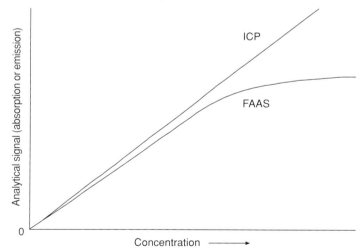

Figure 4 Comparing ICP and Flame Atomic Absorption Spectrometry (FAAS) calibration curves

Instrumentation

The ICP torch consists of three concentric quartz tubes enclosed within an induction coil, which is usually made of copper. The induction coil surrounds the top of the torch and is connected to a radiofrequency (RF) generator (Figure 5). Argon gas flows through all three tubes.

RS•C

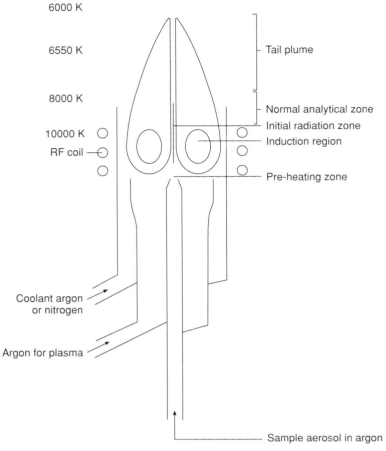

6000 K

6550 K   ⌐ Tail plume

8000 K   ⊢ Normal analytical zone
                                                ⊢ Initial radiation zone
10000 K ○   ○   — Induction region
RF coil —○   ○
         ○   ○   — Pre-heating zone

Coolant argon
or nitrogen

Argon for plasma

                                                Sample aerosol in argon

Figure 5 An ICP torch

┌───┐
│ **Radiofrequency generators**   **Box 2** │
│   │
│ There are two main types of RF generator. A crystal-controlled generator   │
│ uses a piezoelectric quartz crystal to produce an alternating current (AC) │
│ signal that is amplified by the generator before being applied to the   │
│ induction coil.   │
│   │
│ A free-running generator operates at an AC frequency that depends both on │
│ the generator circuitry and the conditions within the plasma. This has the │
│ advantage that the output power tends to remain constant despite changes   │
│ within the plasma. Free-running generators are usually smaller and cheaper │
│ than crystal-controlled generators, and are generally the preferred option.│
└───┘

The sample is introduced as an aerosol into a stream of argon in the central tube. Radiofrequency power (700–1500 W) is applied to the coil, and both electric and magnetic fields are set up in the region at the top of the ICP torch. The acceleration of electrons within the magnetic field produces inductive coupling – a process of adding energy to the argon. Electrons are stripped from the argon atoms when a spark is applied. Collisions between electrons and argon atoms produce a chain reaction that leads to the formation of the plasma.

RS•C

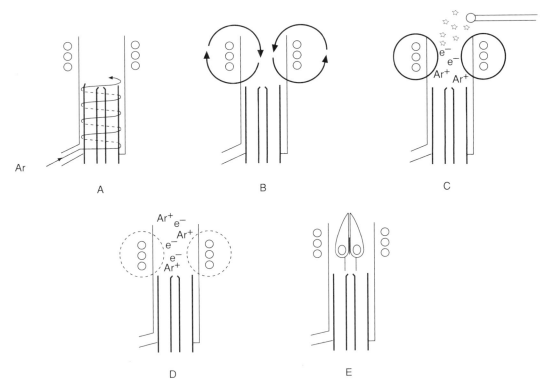

Figure 6 A cross-section of an ICP torch and induction coil showing the ignition sequence
A – Argon gas swirls through
B – RF energy is applied to the induction coil
C – A spark causes some argon to ionise and form free electrons
D – The electrons are accelerated by the RF fields, causing further ionisation so that a plasma forms
E – The flow from the nebuliser, carrying the aerosol of the sample, drives a channel through the plasma

The plasma is made of ions and free atoms together with a high density of electrons. An electric current passes through the induction coil continuously, creating the magnetic field that makes electrons in the argon travel in circular paths. Argon flowing between the inner and outer tubes sustains the plasma and lifts it clear of the quartz tubes, thereby minimising damage to the tubes.

The coolant and auxiliary gas flows enter the torch and produce a vortex flow with a 'weak spot' in the base of the plasma. This has two important effects.

It enables the sample argon stream to drive a channel through the centre of the fireball so that a toroidal-shaped plasma results. The sample has no direct contact with the hottest part of the plasma but is indirectly heated in a very narrow stream within it.

The flow produces circulation of the plasma gases, ensuring thorough mixing. Since most of the energy resulting from induction is concentrated in the outer region of the plasma, its centre is at a lower temperature than other areas. Recirculation allows the inner core to be heated by convection as well as by conduction and radiation. Melting of the quartz tubes is avoided, even at these high temperatures by a flow of argon through the outer tube which cools the rest of the system.

RS•C

Nebulisers   **Box 3**

The nebuliser system is similar to that used in AAS, except that the bores of the nebuliser in the ICP torch are much finer, consequently care must be taken to prevent blockage. High speed argon flowing past the tip of the capillary creates a low pressure region into which the solution is aspirated as an aerosol.

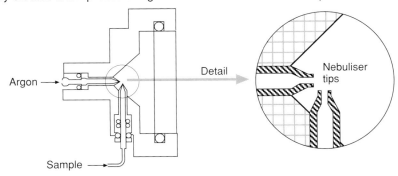

Figure 7 A crossed flow nebuliser

A variation known as a Babington nebuliser can be used for slurries and viscous liquids. In this design the sample flows over a spherical surface and is then converted into an aerosol as a high speed stream of argon is forced through a tiny hole in the surface.

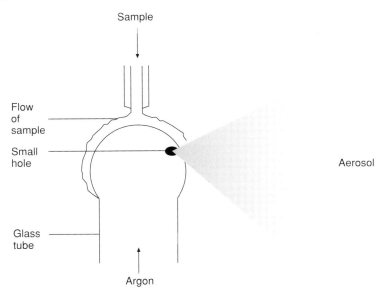

Figure 8 A Babington nebuliser

Detection

The basic detection system for ICP–AES consists of:

■ collimating lenses or mirrors to focus the emitted light through an entrance slit on to the diffraction grating, which disperses the emitted light according to wavelength;

■ a means of separating individual wavelengths; and

■ photomultiplier tubes (PMTs).

RS•C

This system can be adapted for either simultaneous and/or sequential analysis, though some instruments incorporate both types. Speed is the advantage of simultaneous analysis, since the identity and concentrations of a number of elements can be detected and measured at the same time. Each emission line can be observed during the whole sample introduction period. In this system fixed PMTs are set up and positioned with respect to the diffraction grating to receive specific wavelengths – *ie* one PMT for each element. In sequential analysis the light is separated by a monochromator into different wavelengths which are detected in scanning mode by one PMT. With polychromators, the system is normally aligned for specific emission lines making spectral interferences difficult to allow for.

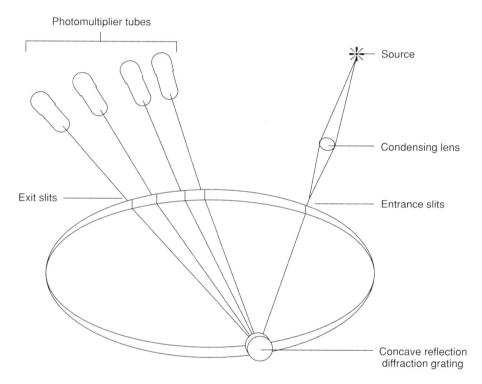

Figure 9 A polychromator system
Photomultiplier tubes (PMTs) are set to specific wavelengths – *eg* to monitor elements that appear consistently in an online process (for simultaneous analysis)

When the sample type is relatively constant and the speed of analysis is important, the polychromator system is advantageous, whereas in a general laboratory environment, the monochromator system is more advantageous because it is usually designed to scan in the important 190–450 nm region.

One advance in the detection of electromagnetic radiation is the use of diode array detectors (DADs) which can respond to a number of different wavelengths simultaneously. Diode array detectors consist of silicon chips incorporating up to 200 pairs of photodiodes and capacitors. Each photodiode is 0.05 mm x 0.5 mm in area and is sensitive to a broad spectrum of radiation, therefore a wide range of wavelengths can be monitored without the need for mechanical scanning. The photodiode arrays are aligned to receive dispersed radiation from a monochromator. The current generated by each photodiode is proportional to the intensity of the radiation received. The signals are digitised, stored and displayed when needed.

For precise measurements (± 2%), the concentration should be at least 50 times greater than the lower detection limit and should be well within the linear dynamic range

RS•C

(LDR) of a chosen emission line. Emission lines with long LDRs make calibration simpler and involve less sample dilution.

Limitations of ICP–AES

Inductively coupled plasma – atomic emission spectrometry instruments are expensive and the large volumes of argon used mean high running costs.

During ICP-AES a large number of emission lines are produced and, if the matrix is complex, this can result in line overlap and considerable spectral interference. What is being measured is the analyte which is contained within the formulation of the sample known as the matrix.

Changing plasma conditions for given elemental wavelengths can overcome line overlap problems. Software developments now mean that, for a particular method, a full set of parameters can be stored and potential problems anticipated.

The energy available in ICP–AES is insufficient to excite fluorine, chlorine and the noble gases. Artificial elements (elements 103–111) are not analysed because of radioactive contamination.

Developments in ICP–AES

Axial plasma viewing
Some of the most modern atomic emission spectrometers view the light being emitted from the analyte axially rather than radially in relation to the plasma.

Viewing axially means looking down the central channel of the plasma, thereby collecting all the emissions over its entire length. The advantage is that the effective path length increases. This leads to improved sensitivity and lower detection limits. The viewing height in the radial system can be altered by looking at that part of the plasma where the atoms and the ions are all in a high energy state.

However, in the axial system the cooler tail plume of the plasma, which includes atoms in the ground state, is also viewed. This can cause significant self-absorption giving rise to non-linear calibration curves but this problem can be avoided by blowing the tail plume out of the optical path using a shear gas, thereby eliminating the zone containing the ground state atoms.

Applications

On-line analysis
Inductively coupled plasma – atomic emission spectrometry (ICP–AES) is being used increasingly for on-line process control, because of its potential for multi-element analysis, its detection limits in parts per billion and its ease of use. Recent legislation on the regulation of waste products and the demands for product quality call for an atomic spectrometric method that can detect certain metals in trace quantities. Two areas in which ICP-AES is useful are the food industry and the industrial monitoring of water effluent. In the food industry, certain trace metals, such as copper, act as catalysts in degrading food and other products, hence continuous monitoring of food during the production phase ensures better quality and longer shelf-life. On-line monitoring of waste water now enables industries to demonstrate to regulatory authorities that they can monitor their effluent and remain within discharge limits.

In process control ICP (PC–ICP), material is routed to an overflow sampler where ICP–AES monitors the output. Appropriate software sets the frequency of monitoring – ie the interval between each measurement, and the flow rate.

RS•C

Process control-ICP is used in power stations and in the aircraft, food and detergent industries. It can monitor stack gases and alloy production as well as liquid effluent.

Assessing trace impurities in silicon chip manufacture

In silicon chip manufacture, semiconductor alloys have to be deposited on a suitable substrate so that a high level of purity is maintained and the layers of crystals are orientated correctly. A vapour phase technique known as organometallic vapour phase epitaxy is used. For example, gallium arsenide (GaAs) semiconductors can be made by decomposing trimethyl gallium vapour, $(CH_3)_3Ga$, in the presence of arsine (AsH_3).

To produce efficient semiconductor devices, trace levels of impurities such as zinc and tin have to be monitored and kept below 1 part per million (ppm). The purified organometallic vapour is injected directly into the spectrometer to detect these trace impurity levels.

Surface coatings

Organisations such as the Paint Research Association use ICP–AES to check the levels of toxic metals in the paints and coatings of children's toys. Regulations such as maximum permissible limits and testing protocols are incorporated in those of the European Committee for Standardisation. Samples of the coating are shaken and dissolved in 50 times their mass of 0.07 mol dm^{-3} hydrochloric acid solution and are then analysed using ICP–AES, particularly for levels of antimony, arsenic, barium, cadmium, chromium, lead, mercury and selenium.

Safe goulash

Paprika is used as a spice in goulash but has historically been adulterated by profiteers. For example, red lead (Pb_3O_4) has been added to some Hungarian paprika to enhance its colour and to add to its mass, resulting in some deaths and serious illnesses. ICP–AES has been used to distinguish between the authentic paprika and the adulterated samples.

Wear in engine oils

Analysing engine oils is a valuable technique for determining wear and tear in engines in the car and aviation industries. When an engine starts to wear after extended use, metal particles from the engine are often found in the oil.

Samples are most conveniently analysed in solution in ICP–AES, and introducing viscous oils causes problems for efficient nebulisation. Engine oils are therefore diluted in an organic solvent which has low viscosity, volatility and toxicity.

Question

The following letter appeared in the journal *Analysis Europa*. You may need to refer to the section on sample preparation in the section on atomic absorption spectrometry to help you with the answer.

'I am doing a project to analyse trace elements in steel by ICP, but I am having problems dissolving the lump of steel. I cannot use a sulfuric acid/phosphoric acid mixture because phosphorus is one of the elements I need to determine in steel. Does anyone have any idea which acid mixture I should use? Also I do not want iron(III) oxide to form.'

RS•C

Answer

The response to the letter was:

'You need aqua regia – a 3:1 hydrochloric acid: nitric acid mix. It is one acid mixture that would probably work. Others include nitric acid/chloric(VII) acid digestion procedures, which should only be done using the correct apparatus, including a wash-down fume hood.'

Inductively coupled plasma – mass spectrometry (ICP–MS)

The high temperatures of the plasma used in ICP techniques ensure atomisation and significant ionisation of a sample, so that free ions can be run through a mass spectrometer for analysis. Linking the ICP torch to a mass spectrometer is known as ICP–MS.

A quadrupole mass spectrometer   **Box 4**

The ICP torch is connected to a quadrupole mass spectrometer. Like other mass spectrometers it separates ions according to their mass-to-charge (m/z) ratio, and produces a mass spectrum. The quadrupole mass spectrometer has several important characteristics:

- it is very compact because there is no magnet (it can be used as a bench-top instrument);

- it requires lower voltages than other mass spectrometers; and

- it scans very rapidly over a wide range of masses.

However, the quadrupole instrument is not as sensitive as magnetic instruments.

A quadrupole mass spectrometer consists of four parallel rods, opposite pairs of which are connected electrically. When direct current (DC) and radiofrequency (RF) voltages are applied across the rods, only those ions with a specific m/z ratio follow a stable path to the detector. The rest follow an unstable path, colliding with the rods. By altering the ratios of the DC and RF fields, ions with different m/z values take the stable path to the detector.

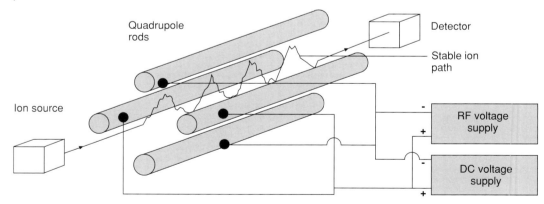

Figure 10 A schematic of a quadropole mass spectrometer

This coupling of ICP with a mass spectrometer (ICP–MS) is particularly useful. A mass spectrum gives isotopic information, not available through ICP–AES, and atomisation and ionisation through ICP means that a rich source of ions is available for the mass spectrometer. Essentially, the system can be seen as a source and detector. The ICP is

RS•C

the source of ions and the mass spectrometer is the detector by mass. The mass spectrometer is both qualitative and quantitative, identifying isotopes from mass numbers of the ion peaks and by measuring the peak height. Peak area is measured if the spectrometer is equally sensitive to all elements in the mixture. Figure 11 shows a typical spectrum.

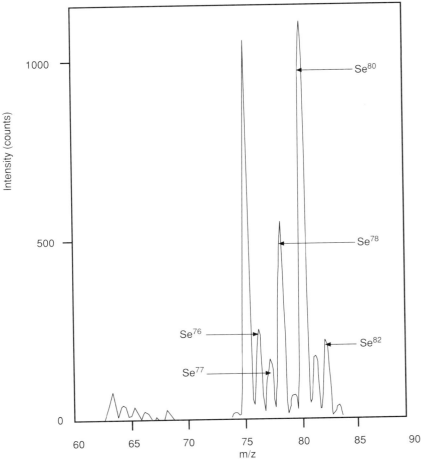

Figure 11 Selenium in aqueous solution, detected using low pressure helium ICP–MS

The detection limits in ICP–MS are better than in ICP–AES, because ions can be detected at lower levels and with greater precision than photons.

The interface between the ICP torch and the mass spectrometer is crucial because ICP operates at atmospheric pressure while the mass spectrometer is kept under vacuum. The interface consists of a series of nickel cones with small bore holes, so ions can be transmitted from the plasma with a progressive reduction in pressure between cones.

The hotter regions of the plasma are used as the source because there is a greater concentration of ions here. However, there is interference from background molecular ions. The torch is positioned horizontally, end-on to a water-cooled sampling cone; the aperture diameter is usually between 0.5 mm and 1.5 mm.

After being carried through a series of chambers held at progressively lower pressures, the ions pass through the interface. Neutral species are removed by vacuum pumps.

Laser ablation (vaporisation) vaporises solid samples that are fed directly into the plasma. However, it is difficult to standardise solid samples. For qualitative analyses, laser ablation is an attractive alternative to dissolving intractable solids like rocks. Many

RS•C

geological samples are digested with hydrofluoric acid (HF) to dissolve silicates. This is done in polytetrafluoroethene (PTFE) containers because HF attacks glassware.

Interferences in ICP–MS

Spectral

Spectral interferences occur when one or more species have similar masses to the analyte and the mass spectrometer cannot distinguish between them. For example, ^{58}Ni and ^{58}Fe, ^{40}Ar and ^{40}Ca. Using mixed gases in the plasma, such as adding nitrogen to argon, helps to reduce interference.

Refractory oxides

The survival of refractory oxides – *eg* oxides of metals such as antimony, aluminium and molybdenum – through the plasma can be remedied by low nebuliser flow, higher power, and greater spacing between load coil and sampler.

Limitations of ICP–MS

In a mass spectrometer there can be molecular interference with combinations such as ArO^{+} (M_r 56) interfering with the detection of iron (A_r 56) – *ie* they are isobaric. High concentrations of dissolved solids can result in clogging of the skimmer cones. A build up of deposits in the spectrometer causes loss of sensitivity and increases maintenance costs. To avoid this problem, many samples have to be diluted up to 1000 times which raises the detection limits considerably, bringing a few of them above those of ICP-AES. There are also matrix effects, which cause problems in the mass spectrometer. Consequently, the analyte sometimes needs to be separated from the matrix, resulting in time consuming chromatographic separations.

One final consideration is the context in which ICP–MS can be used. Although the technique can be used to detect many elements down to parts per trillion (ng dm^{-3}), the reagents and glassware have to be clean enough to match these detection limits. This is often difficult to achieve. However, where the matrix effects are relatively unimportant, or where extremely low detection limits are needed, ICP–MS has a role. The heavy metals investigated in rocks have atomic masses greater than those reached by argon and oxygen (ArO^{+}), which is isobaric with iron. In the nuclear and water industries there are few matrix problems when detecting trace contaminants.

Applications

The case of the paint spot murder

ICP-MS has helped to convict a murderer. In 1991 and 1992 two men in Lincoln were bludgeoned to death with an axe. By chance the axe was discovered in the undergrowth of an island in a boating lake. It had apparently been painted between the two murders, and police thought that identifying the source of the paint might give them a lead. Investigations were carried out at the Paint Research Association. Two pieces of information emerged. First, from the ICP–MS analysis, the zinc in the paint was thought to be recycled because of the presence of other metals, and secondly, the binder that helps the paint stick to surfaces was Epoxyester D4. Only two firms in Europe manufacture paint that contains both recycled zinc and Epoxyester D4, and only one of these exports to the UK. This firm produces a spray-on paint called Rustoleum 2185, 'used mainly for touching up scratched bare metal'. ICP–MS tests on samples provided by the paint manufacturers confirmed that it was identical to the paint on the axe head.

RS•C

Having pinpointed the paint and the suppliers in the UK, police were able to trace Rustoleum 2185 that had been sold from an engineering tool merchant in Lincoln. It was found that a pot of the paint had been sold to a suspect. From this point, the remaining loose ends were tied up, and the suspect was convicted.

Analysing minerals

When interfaced with a powerful microscope and a laser, ICP–MS provides accurate chemical information on the spatial distribution of minerals in a rock.

By using a microscope, a particular section of the slide can be moved into focus and a small region of the surface can be evaporated using a laser. The vaporised fragment passes to the plasma of the torch where ionised atoms are drawn through the nickel skimmer cones into the mass spectrometer. This gives information about the elements present in that tiny piece of rock, and their distribution in neighbouring crystals.

For example burning Spanish coal produces large amounts of oxides of sulfur (associated with acid rain), and associated with the sulfur-rich minerals were potentially toxic heavy metals. The aim of the analysis was to identify the location of the heavy metals in the coal, which were found to associate with the smallest sulfur-rich granules. These are too small to remove, hence no physicochemical cleaning process could dispose of the heavy metals efficiently.

The analysis of illicit heroin sources

Heroin can be characterised by analysing component trace elements. Following the distribution path of heroin – so that prosecutions can be made, requires discovering the route of the drug from its natural plant source, through the manufacturing process, to the pusher. The trace inorganic impurities found in a heroin sample indicate its source. ICP–MS provides a 'fingerprint' to locate sources and therefore any routes through which the heroin is distributed.

Jet fuels

Copper and iron are catalysts in jet fuel oxidation. ICP-MS is used to monitor their presence because even at parts per billion (ppb) levels these can be very dangerous.

Beating gold smuggling

In Western Australia the annual cost to the gold-mining industry of stolen gold is around A$20 million. Laser ablation of gold samples in an ICP and detection through mass spectrometry can locate the source of gold that may have been smuggled or stolen from gold quarries. The gold samples have a typical impurity 'fingerprint' depending on their origin. Using ICP–MS means that a library of fingerprints from different sources can be generated and checked against any suspect samples.

The history of ancient bronzes

ICP–MS can indicate changes in methods and technology for casting ancient bronzes. In the early Bronze Age (3000 BC–1500 BC) the prevalent method of making bronzes was by smelting together copper and arsenic minerals. Tin started to be added in the middle Bronze Age

RS•C

onwards and in some places lead was added as a casting material. By the time of the Romans indications are that brass was also being made.

Tracing the trace element

Trace elements such as iron, copper and selenium are a vital part of our diet. They are needed only in trace quantities, but their absence can result in disease. The way in which these elements influence our health depends on a number of factors, including amount, oxidation state, whether they are in organic or inorganic form, and whether they are taken in association with other elements. For example, chromium(III) is beneficial to health whereas chromium(VI) is toxic. High zinc uptake affects copper absorption.

Selenium can be taken up in both organic and inorganic forms such as selenium methionine and sodium selenate (Na_2SeO_4). It is known that organic selenium, but not the inorganic form, has a role in preventing mammary tumours in rats, and a selenium deficiency in sheep feed results in fatty degeneration of coronary arteries. The difference between a healthy daily intake of selenium and a potentially toxic one can be quite small. The recommended daily intake (RDI) of selenium in the UK is 50 µg for adult males, more than 200 µg can produce disturbed selenium metabolism. To understand the effects of selenium intake, we need to know the form of the trace element in the food: what happens to it in the gut after ingestion, digestion and absorption in the blood stream, and the effect on the element's target sites in the human body.

One way of investigating this is to follow selenium uptake in human volunteers by blood sampling, and urine and faeces testing using radiolabelled selenium isotopes. The challenge then is to remove the element from the matrix without changing its chemical form. Selenium uptake has been modelled using dialysis to simulate movement across the gut during digestion. The selenium is then separated from the food matrix of selenium yeast-based supplement tablets (used in the modelling) using ion exchange chromatography and detected by using ICP–MS.

Ancient liquids and precious resources

As raw materials in the surface layers of the earth are gradually being depleted, there is increasing interest in making deeper explorations. Ancient geological fluids concentrated many of the world's valuable resources, such as phosphates and metals, in the deeper layers of rock. Microscopic remains of these fluids, known as 'inclusions', have been trapped in rocks for millions of years and are important pointers to the location of useful deposits. For example, there is a link between the location of major gold deposits and the outgassing of carbon dioxide rich fluids from the Earth's lower crust.

Concentrated saline fluids were involved in the formation of rich copper seams, and liquid hydrocarbon droplets provide information on the formation of oil-water reservoirs.

Laser Ablation ICP–MS (LA–ICP–MS) detects 'inclusions'. An optically-guided ultraviolet laser drills through the rock sample to release the contents of fluid inclusions for analysis by ICP–MS. The sensitivity of the technique is particularly valuable because usually only tiny droplets are available. The chemistry of these droplets can be investigated and linked to the location of mineral deposits.

RS•C

Choosing the method

Future developments and other issues

Some of the detection limits reached by AAS, particularly in hydride generation and graphite furnace techniques, are likely to remain better than any other analytical technique for the foreseeable future. While ICP–AES cannot match the low detection limits of graphite furnace atomic absorption spectrometry (GFAAS) for most elements, its main advantages are speed, few interferences and, of course, multi-element analysis.

However, AAS remains a viable technique. First, it has long been used for validating standards for a range of materials. Secondly, it is a quick and cost-effective method for determining samples containing less than five elements, and thirdly, the detection limits of flame atomic absorption spectrometry (FAAS) are adequate for most industrial purposes.

The instrument and method analysts choose is the one that yields the most reliable and valid data and requires minimum cost and time. Quite often it is a matter of making compromises and judgements. What might appear to be a very sensitive technique, perfectly adapted to the needs of a research department, might not be so good when considered in an industrial context. For example, ICP–MS has the lowest detection limits (for some elements down to parts per trillion (ppt) and better), but the dilution factor needed for detecting light elements makes the limits of detection for ICP–AES in some cases better than ICP–MS. For most quality assurance needs in industry, the low detection limits that can be reached by ICP–MS are not necessary. There are also problems with using ICP–MS to detect iron, because iron has the same m/z ratio as ArO^+ – formed from the plasma gas and air. On a cost-benefit basis, in routine industrial analysis, ICP–AES seems a better option. However, if only one or two elements per sample need to be detected, then AAS is more attractive. Graphite furnace atomic absorption spectrometry (GFAAS) is more sensitive than FAAS, but where detection limits of parts per million (ppm), or just under, suffice, FAAS is a good choice.

When choosing an appropriate technique there is a host of interconnected factors to consider. These are shown in Table 1.

RS•C

Parameters	Flame atomic absorption spectrometry (FAAS)	Graphite furnace atomic absorption spectrometry (GFAAS)	Inductively coupled plasma-atomic emission spectrometry (ICP–AES)	Inductively coupled plasma-mass spectrometry (ICP–MS)
Sample throughput	Fast	Slow	Fast	Fast
Number of elements within a sample	Usually one, but some modern instruments can do up to four at once	As for FAAS	Many	Many
Linear dynamic range (LDR)	Poor	Poor	Good	Very good
Maximum proportion of dissolved solids	> 20%	~3%	2–25% depending on the nebuliser	0.1–0.4%
Costs	Low	Higher than FAAS.	High	Very high
Running costs	Low	Medium	High	Very high
Detection limits	Good for some elements.	Excellent for some elements. Better than FAAS for most.	Good for most elements.	Excellent for most elements.
Precision	0.1–1%	1–5%	0.3–2%	1–3%
Interference	Many chemical interferences	Many chemical interferences	Almost none	Mainly from isotopes
Chemical fuels	Yes (ethyne for the flame)	None	None	None

Table 1 A comparison of atomic spectrometric techniques

Questions

1. Which of the methods that have been described in atomic absorption spectrometry (FAAS, GFAAS, HGAAS, CVAAS, ICP–AES, ICP–MS) would you use for the following analyses?

   a) Detecting mercury at a level of less than 1 mg kg^{-1} in fish products.

   b) Detecting trace levels of cadmium in the blood of pregnant women living in the vicinity of a copper smelter.

   c) Detecting precious metals in rocks.

   d) Detecting lead in dust.

   e) Detecting and determining heavy metal concentrations in effluent water samples from a steel works.

   f) Determining selenium concentrations in blood.

RS•C

2. You have an atomic absorption spectrometer that can be used for flame, electrothermal or hydride generation techniques. Decide which option you would use in the following situations.

a) The clothes of a murder suspect have pieces of plastic clinging to them. These appear to come from a worn car seat where the crime was committed. The car seat fabric is known to contain antimony as a flame retardant in low concentrations.

b) To find out whether copper salts from a nearby copper mine have leached into water supplies.

Answers

1. a) Mercury is determined most sensitively using cold vapour atomic absorption spectrometry (CVAAS).

   b) Trace levels of cadmium are best detected using graphite furnace atomic absorption spectrometry (GFAAS).

   c) Inductively coupled plasma-mass spectrometry (ICP–MS) is the best technique for multi-element analysis and can be used for detecting elements in low concentrations in rocks.

   d) Flame atomic absorption spectrometry (FAAS) is the best technique for detecting lead in dust, however, GFAAS may be a better option depending on the sensitivity needed.

   e) Again ICP–MS is the best technique for multi-element analysis.

   f) Inductively coupled-atomic emission spectrometry (ICP–AES) is adequate for determining selenium concentration in blood, because it is only selenium that is being investigated.

2. a) Either hydride generation atomic absorption spectrometry (HGAAS) or graphite furnace atomic absorption spectrometry (GFAAS) could be used to detect the antimony. HGAAS has lower detection limits, but the plastic sample has to be digested and dissolved before analysis. If all that is required is confirmation of the presence of antimony, then the material could be used in GFAAS without any sample preparation.

   b) The copper is already in solution and FAAS is the best means of analysis here.

RS•C

This page has been intentionally left blank.

RS•C

X-ray fluorescence spectrometry

Introduction

The Tower of London is an unusual place to find modern sophisticated analytical instruments but lodged in a 13th century cell in the Tower, an X-ray fluorescence (XRF) spectrometer is used to determine the elemental composition of ancient armour.

In more conventional locations XRF spectrometers have been used to:

■ determine the elemental composition of the blue soda glass from York Minster, where interpretation of the data indicated three distinct sources of raw materials; and

■ authenticate Chinese porcelains, so fake Chinese pottery can now be identified based on its barium composition.

X-ray fluorescence is particularly useful for analysing archaeological artifacts because it is non-destructive. Other applications of XRF spectrometry are:

■ the quality control of raw materials and chemicals in the paper industry;

■ the on-line analysis of flowing slurries in the cement industry;

■ the analysis of geological materials; and

■ determining the wear and tear in aircraft engines by analysing metals in the fuel.

Although XRF spectrometry is less sensitive – parts per million (ppm) rather than parts per billion (ppb), XRF has one major advantage over atomic absorption (AA) and inductively coupled plasma (ICP) techniques. It can be used to analyse artifacts *in situ*, as well as not requiring specially-prepared solids and solutions. In fact, ICP and XRF techniques can complement each other. For example, X-ray fluorescence can be used to determine the metals in bulk materials in landfill sites, while the leaching of metals from those sites can be analysed to lower limits of detection using ICP techniques.

The basis of XRF (Figure 1) is that elements in a sample emit characteristic X-rays – recorded as peaks on a spectrum – when excited by an X-ray source, or other sources such as electrons, protons or α-radiation. Measuring the wavelength and the intensity of each peak gives information on the identity and quantity of each element present in a sample.

RS•C

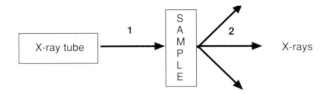

1. Primary X-ray source is incident on the sample.
2. Secondary X-rays of different wavelengths are emitted from the sample.

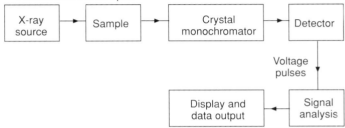

Sequence of components in wavelength dispersive x-ray spectrometer

Figure 1 A schematic representation of XRF

Some advantages of XRF spectrometry

- Samples can be analysed in situ.

- Data are obtained within minutes.

- The data are relatively simple, being a sequence of lines which can be quickly identified.

- There is minimal interference in the spectra.

- Little sample preparation is needed.

- Sample holders can be adapted to analyse samples in a variety of forms – bulk solids, powders, slurries, liquids and even gases.

Principles

X-ray fluorescence spectrometry is a surface technique which involves measuring the wavelengths (or photon energies) of X-rays that are emitted from a fluorescing sample. When X-rays with energies above a certain level interact with atoms the following changes occur (Figure 2).

- An electron is ejected from one of the inner shells (or energy levels) of the sample, leaving a 'hole'.

- There is a rearrangement within the atom so that an electron falls from an outer shell (or energy level) to fill the 'hole'.

- X-rays are emitted – their wavelengths being characteristic of the elements present in the sample. The wavelengths of these secondary X-rays are longer than the wavelengths of the incident primary radiation – *ie* the sample fluoresces.

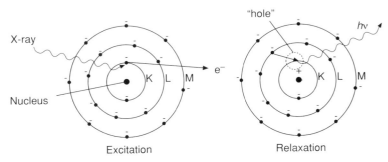

Figure 2 A schematic representation of an X-ray fluorescence process
(K, *n*=1; L, *n*=2; M, *n*=3)

The rules that govern the electron transitions (Figure 3) are based on the selection rules of quantum mechanics.

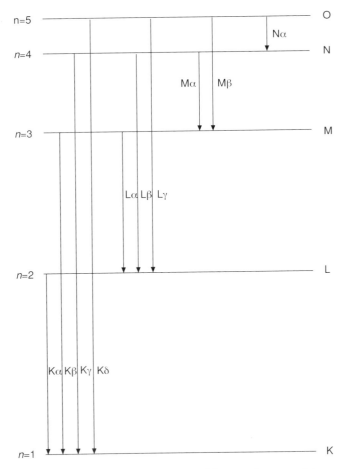

Figure 3 Diagram summarising transitions and nomenclature

The principal quantum number must change by at least one – *ie* an electron cannot be replaced by an electron within a shell of the same energy level. For example, fluorescence may occur when an electron falls from a higher to a lower energy level such that its principal quantum number decreases by at least one – *eg* from $n=2$ to $n=1$, or $n=4$ to $n=2$.

RS•C

The orbital quantum number must also change by only one and other transitions are forbidden – *ie* s to s replacements are forbidden transitions, but p to s, s to p and d to p replacements are allowed.

Explaining electronic transitions   Box 1

The principal quantum shells are designated K, L, M, N, O *etc* in XRF notation whereas the usual representation is $n = 1$, 2, 3 etc.

For

K, $n = 1$,
L, $n = 2$
M, $n = 3$ *etc*

K, L, M, N *etc* is the shell where the electron comes to 'rest'.
α signifies that the electron has fallen only one shell *eg* from the L shell to the K shell. β signifies that it has fallen two shells and γ three shells.
The numerals following the α or any other Greek symbol depend on the orbital quantum number of the electron in both its initial and final resting states *ie* one shell.

There are some broad generalisations that help to identify the lines in a spectrum.

The $K\alpha_1$ and $K\alpha_2$ lines are so close together that they are often undifferentiated.

$K\beta$ lines are about 1/5 as intense as the $K\alpha_1$ line

The $K\beta$ line has a shorter wavelength than the α lines because the electronic fall in energy is greater.

Transitions to the K lines have the shortest wavelengths, followed by L, M, N... *etc*.

The limitations of XRF analysis

When X-rays interact with the atoms of an element two electronic changes may occur. The first has been discussed, when an electron is ejected from an inner shell and the 'hole' is filled by an electron from a higher energy level. This results in XRF. However, there is a competing change known as Auger electron emission (Figure 4), which does not result in fluorescence. The first two stages are the same as XRF – the atom is excited and electronic rearrangement occurs - but the escaping X-ray photon is absorbed when it interacts with an outer electron of the atom. This electron is emitted instead of the photon. As a result the atom becomes doubly ionised – *ie* two electrons are lost.

RS•C

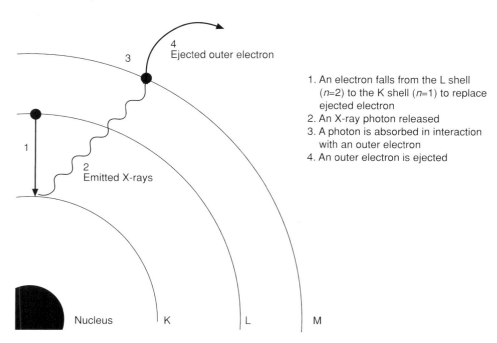

1. An electron falls from the L shell (*n*=2) to the K shell (*n*=1) to replace ejected electron
2. An X-ray photon released
3. A photon is absorbed in interaction with an outer electron
4. An outer electron is ejected

Figure 4 Auger electron emission

Auger emission is common in elements with low atomic numbers because the nuclear charge in such atoms is relatively small so that it is comparatively easy to eject electrons from the lowest energy levels.

X-ray fluorescence spectrometry is not used to determine elements below atomic number 4 because of the competing effect of Auger electron emission.

Auger electron spectroscopy   **Box 2**

The Auger effect is used to detect and quantify light atoms by measuring the energies and relative intensities of the ejected electrons. Most applications are in surface studies such as studying the concentrations of elements with low atomic numbers in the seams caused by metal fractures.

Instrumentation

The most commonly used instrument in XRF analysis is the wavelength dispersive XRF (WDXRF) spectrometer (Figure 5). Its basic components are:

- an X-ray source, usually an X-ray tube;

- a sample holder;

- collimators for eliminating stray radiation;

- a crystal for diffracting and dispersing the X-ray beams emitted by the fluorescing sample; and

- a detector which converts the X-ray photons into voltage pulses.

RS•C

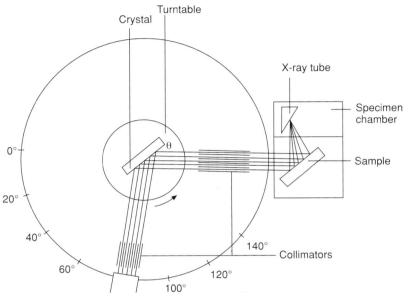

Figure 5 The layout of a typical wavelength dispersive spectrometer

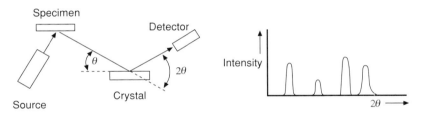

Figure 6 The principle of a wavelength dispersive spectrometer, and a generalised spectrograph. A simple wavelength can be selected by selecting the equivalent value of 2θ

The source provides an X-ray beam that irradiates the sample. When the primary X-rays have energies greater then characteristic energies for elements in the sample, secondary X-rays of different wavelengths are emitted. A crystal, acting as a monochromator, diffracts the radiation according to the Bragg equation – $n\lambda = 2d\sin\theta$ – (see Figure 7 and *X-ray crystallography* section), dispersing the various wavelengths in the X-ray beam. Diffraction occurs only when the distance travelled by waves in successive planes of the crystal differs by a whole number of wavelengths.

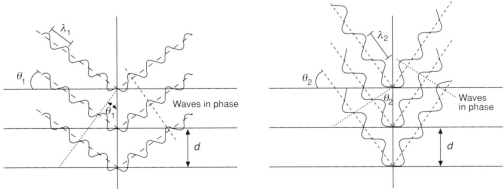

Figure 7 In a crystal with spacing *d*, waves of wavelength λ_1 at angle θ_1 are different in wavelength to those diffracted at angle θ_2, hence there is dispersion between the waves

RS•C

By changing θ – ie rotating the crystal – different wavelengths are diffracted. The data obtained by the detector are used to identify the elements in the sample as each peak – known as a 'line' – is recorded at 2θ. Curved crystals are used in modern spectrometers to improve resolution and to direct the radiation more efficiently towards the detector.

Choosing the crystal for diffraction   **Box 3**

The separation between X-ray peaks depends on the relationship between wavelength and the d-spacings of the crystal. A number of different crystals are used to cover the full measurable range. Some of the more popular crystals are lithium fluoride and germanium. For example, germanium covers the range of wavelengths suitable for determining chlorine, phosphorus and sulfur. A recent development is the manufacture of multilayer crystals, with extremely small d-spacings, for determining elements with low atomic numbers such as boron and carbon.

Sample preparation

Compared with atomic absorption and ICP techniques, XRF is much more versatile because it can handle many types of sample. Instruments can even be adapted to measure irregular solids. However, bulky solid samples are usually cut down and shaped in cylinders about 2–5 cm in diameter and 1–3 cm in length. Where solids cannot be shaped they are milled into powders and then pelleted into cylinders. The advantage of milling and pelleting is that a solid can be homogenised for quantification. Commercial standard powders can be added as standard additions where different amounts of the analyte standard are added to the sample to overcome matrix effects in calibration. (A case study of a standard addition is described later.) Standard additions are also covered in the section on Atomic absorption spectrometry (AAS). For samples like dry sands, where there is low cohesion between particles, a binder, such as cellulose, is mixed with the sample.

The main problem with sample preparation of bulk solids is the roughness of the surface. Solids, such as ceramics, with normally smooth surfaces need no surface preparation. However, the surfaces of most other solids have tiny grooves, usually only a few micrometres thick, and these can affect peak intensity (Figure 8). Some parts of an uneven surface can prevent other parts from receiving radiation, hence peak intensity decreases with roughness.

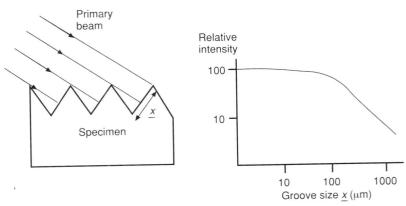

Figure 8 How surface roughness can influence line intensity

Where possible, powders, or solids with small particle sizes, are preferred to bulk solids. Since XRF is primarily a surface technique it is best to have as large a surface area as possible available to increase sensitivity – the smaller the particle size, the greater the intensity of the lines.

RS•C

Samples of geological materials and many other solid samples are often prepared for quantitative measurement by fusing them with lithium tetraborate ($Li_2B_4O_7$). This gives a homogeneous solid solution.

Solutions of samples are also determined by XRF, and prepared in the same way as those for atomic absorption and ICP techniques.

Quantitative measurements

Quantitative measurements can be made by fixing the detector at a value of 2θ. The detector receives X-rays of a particular wavelength and records the number of counts in a pre-set time. Quantitative measurements require careful calibration. Standards are used and these are carefully matched with the unknown sample for matrix effects and estimated concentration. The matrix comprises all the chemical components present in the sample other than the element – analyte – being measured. Bulk solids should have the same surface finish and powders should have the same packing density and particle size distribution. For example, to determine the concentration of nickel in steel rods where the estimated concentration of the nickel is 2.5% by mass, similar steel rods are analysed which have accurately known concentrations of nickel between 1% and 7%. A particular line, such as the Ni K_α line, is then chosen. The method used to determine the percentage mass of nickel in a sample is outlined in the case study.

Advanced software can now compute algorithms for matrix interferences. Where elements have very different sensitivities – eg sodium has a low sensitivity in XRF compared to calcium – peaks of different height show for the same amounts of different elements. Computers can now compensate for this.

A case study of a standard addition

The manager of a steel production company sent a sample of low-grade iron ore for assessment. The sample was finely ground and divided into roughly equal amounts. Known amounts of iron(III) oxide were added to three of the samples and mixed. The samples were pelleted under identical conditions and placed in an X-ray spectrometer set on the Fe K_α first order line. The following results were obtained for each of the samples.

Sample	Mass of ore/g	Mass of added iron(III) oxide/g	Counts/10 s
1	5.0123	0.0000	9213
2	5.0231	0.5614	11998
3	4.9937	1.2504	14804
4	5.0021	2.1502	17715

A graph is drawn of percentage added iron(III) oxide against count number. Extrapolating the graph back to the x-axis gives the percentage of iron(III) oxide originally in the sample. The percentage mass of iron in the sample can then be calculated.

The percentage added iron(III) oxide for sample 2 can be calculated as follows:

$$
\begin{aligned}
\text{Percentage added } Fe_2O_3 \ = \ & \text{mass of iron(III) oxide added/} \\
& \text{total mass of sample x 100} \\
= \ & 0.5614/(5.0231 + 0.5614) \text{ x } 100 \\
= \ & 10.05\%
\end{aligned}
$$

This percentage figure is then plotted against the count number, 11998. Similar calculations performed for the other two standard additions enable a graph to be

RS•C

drawn and the percentage of iron(III) oxide in the sample calculated to be 32.5 %
(Figure 9).

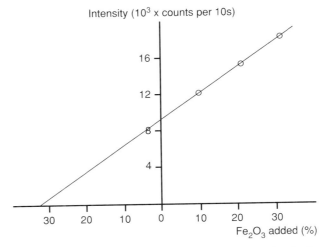

Figure 9 A standard addition calibration curve for Fe$_2$O$_3$
(the original sample contains 32.5 % by mass of iron oxide)
(Adapted with permission from C. Whiston, *X-ray Methods*,
Chichester: John Wiley & Sons on behalf of ACOL, 1987.)

The percentage of iron in the original sample can then be calculated.

Detection

The main detectors used are gas-filled proportional counters (Box 4) or scintillation
counters in conjunction with phosphors. The signal is passed via a cable to a control
unit where the result is displayed.

RS•C

Gas-filled proportional counters   Box 4

Gas-filled proportional counters (Figure 10) consist of an earthed cylindrical tube *ca* 2 cm in diameter containing argon gas, with a thin wire as the anode running along its radial axis. A potential of about 1500 V is applied to the wire.

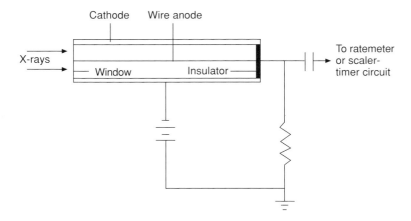

Figure 10 A gas-filled proportional counter

X-ray photons entering the detector through the mica window cause ionisation of the argon gas, producing positively charged argon ions and electrons. The electrons move towards the wire (the anode), accelerating as they approach, and gaining enough energy to further ionise the argon atoms which become doubly charged. Amplification of the numbers of electrons produces a build up of charge at the capacitor. The voltage change across the capacitor is proportional to the energy of the incident X-ray photon. This type of detector works best for wavelengths greater than 0.15 nm.

Scintillation counters

The gas-filled counter is insensitive to shorter wavelengths, so a scintillation counter is used. This counter consists of a photomultiplier and a scintillator or phosphor, usually a single crystal of sodium iodide doped with thallium. When X-ray photons interact with the phosphor, secondary photons are released. The number of photons released is proportional to the energy of the incident X-ray photons. The photomultiplier acts as a transducer, converting the energy from the light flashes into electrical pulses, which are amplified and counted.

RS•C

Sequential and simultaneous spectrometers   Box 5

Modern wavelength dispersive spectrometers are automated and interfaced with computers programmed to the relevant parameters of the analysis, such as crystal used, range of wavelengths scanned and X-ray tube current. They are either sequential or simultaneous spectrometers. Sequential spectrometers (Figure 11) record in turn each wavelength diffracted by the crystal from a fluorescing sample.

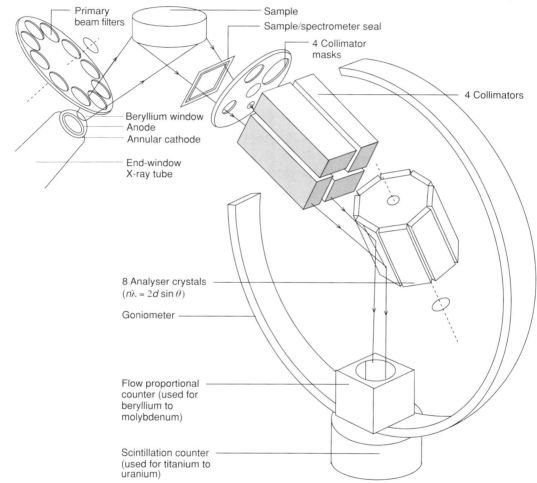

Primary beam filters
Sample
Sample/spectrometer seal
4 Collimator masks
4 Collimators
Beryllium window
Anode
Annular cathode
End-window X-ray tube
8 Analyser crystals ($n\lambda = 2d \sin\theta$)
Goniometer
Flow proportional counter (used for beryllium to molybdenum)
Scintillation counter (used for titanium to uranium)

Figure 11 Diagram of modern sequential spectrometer

Simultaneous wavelength dispersive spectrometers are designed to record the X-rays emitted from specified elements. They consist of as many as 12 single channels arranged around the X-ray tube and sample. Each channel is fixed at a set 2θ angle in relation to the sample and analysing crystal, to detect designated elements.

Advantages of sequential spectrometers
- They are compact.
- They are cheaper than other spectrometers because they have fewer components.
- They are more flexible because they can theoretically detect any element with atomic number greater than 4, whereas simultaneous spectrometers are programmed to detect specific elements.

Advantages of simultaneous spectrometers
- They have no moving parts and so are less likely to develop faults.
- They are suited to online analyses for quality control processes – *eg* cement manufacture, where any variations outside pre-set limits can be quickly detected.
- They have higher sensitivity for trace analysis, where two channels can be set to the same wavelengths and their outputs combined.

RS•C

RS•C

Data display for WDXRF spectrometers   **Box 6**

In wavelength dispersive instruments, the spectrum is a plot of 2θ against intensity. From Bragg's law, for a particular crystal of constant d-spacings, the value of the wavelength is proportional to 2θ. Using modern software, each peak can be labelled with the name of the element and the electronic transition.

Energy dispersive XRF (EDXRF)

Energy dispersive XRF spectrometers (Figure 12) separate the characteristic X-rays emerging from a fluorescing sample according to their photon energies.

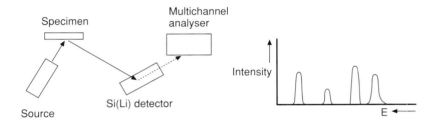

Figure 12 The principle of EDXRF. Selection of a single energy level is achieved by selection of an appropriate range of channels on the multichannel analyser

There is no crystal monochromator so the detector receives all the secondary wavelengths at once. For each X-ray photon incident on the detector, a pulse of electric current is produced which is proportional to the photon energy. The detector output is amplified, and is then subjected to pulse analysis to separate the pulses on the basis of photon energies.

The versatility of this spectrometer is due to the lithium-drifted silicon detector, and powerful microcomputers that decode the information according to the user's needs. The detector is cooled in liquid nitrogen to reduce electronic noise – mainly from the mobility of the lithium atoms, and to enhance resolution.

Energy dispersive spectrometers are much smaller and compact than WDXRF spectrometers because they do not need a crystal monochromator.

RS•C

Comparison	WDXRF	EDXRF
Arrangement of components	The presence of a crystal and a drive mechanism means that these instruments are bulkier than ED spectrometers.	These instruments have no crystal, therefore the instrument is compact and cheap. The sample is placed close to the detector.
X-ray tubes	X-ray tubes operate at higher voltages. Thick windows are needed for safety. Tubes are expensive and need to be replaced more often.	The proximity of the sample to the detector means that a less intense X-ray source is needed. Cheap, low power X-ray tubes can be used.
Speed of analysis	The detector is moved in relation to the sample. Collecting the X-rays is sequential and relatively slow.	All photons are detected simultaneously, so qualitative analysis is rapid.
Sensitivity	Sensitivity for small peaks can be enhanced using long count times, because only radiation of one wavelength at a time enters the detector. Increased sensitivity is more suitable for quantitative analysis.	Low sensitivity for weak lines because the strongest line determines the count rate. This makes it less suitable for quantitative analysis.
Extras	Crystals may need to be changed for detecting certain elements.	Liquid nitrogen for cooling the detector.

Table 1 The comparison between wavelength dispersive and energy dispersive techniques

Other XRF techniques

There are a number of variants on both wave dispersive and energy dispersive instruments. The differences exist primarily in the means of exciting the sample. A specific application of XRF – scanning electron microscopes (SEMs) – is described in *Modern chemical techniques*, London: Royal Society of Chemistry, 1992, (p160).

Total reflection XRF (TRXRF)
In conventional energy dispersive instruments the sample has to be placed on a support material. This produces high background 'noise' due to scattering of the X-rays, adversely affecting the detection limits and the sensitivity of the technique. To achieve lower detection limits, a method has to have the incident X-ray beam exciting the sample but not the support material. Figure 13 shows the arrangement of a TRXRF instrument.

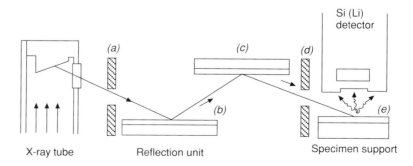

Figure 13 The arrangement for TRXRF: *(a)* **first aperture** *(b)* **first reflection unit** *(c)* **second reflection unit** *(d)* **second aperture; and** *(e)* **the specimen**

RS•C

To achieve this, the sample is prepared as a thin film on an optically flat surface, usually a quartz plate. A pair of reflectors guide the beam coming from the X-ray tube onto the sample at a very low glancing angle. The angle of incidence must be less than the critical angle so that total reflection occurs. The critical angle depends on the properties of the support. (Figure 14).

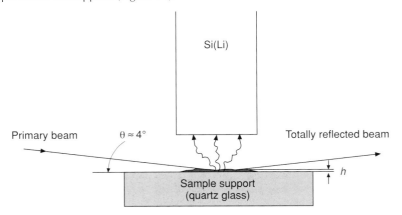

Figure 14 Diagram to show instrument set-up for TXRRF

Any fluorescence from the support material is very low because the X-rays penetrate only the surface of the sample. As a result there is very low background noise and increased sensitivity. Elements can be detected down to picogram (10^{-12} g) levels or concentrations less than parts per billion (10^{-9}, ppb), compared with parts per million (10^{-6}, ppm) detection limits for conventional XRF spectrometers.

Total reflection X-ray fluorescence is used for analysing rain and river waters where elements are often present in trace amounts. It is also used to detect surface contamination of wafers used in the semiconductor industry.

Particle-induced X-ray emission analysis (PIXE)
This is a comparatively rare and expensive technique in which a beam of fast protons, rather than X-rays, is directed towards the sample. The energy of the proton beam is between 1–4 MeV. The system contains:

■ proton accelerators such as van de Graaf generators;

■ an energy-diffusing magnetic deflection field;

■ an electrostatic lens for focusing the beam;

■ a high vacuum target for the sample; and

■ an energy dispersive detector.

The protons irradiate the sample, ejecting inner shell electrons. Characteristic X-rays are emitted as the holes left by this irradiation are filled with electrons from outer shells.

The advantage of PIXE is that it generates only a small amount of background noise and can be used for very small samples. It can monitor levels of heavy metals in aerosol particulates and detect trace elements in wine. For example, PIXE was used to determine heavy metal concentrations in the air over Siberia. This is an area where there is very little heavy industry so any measurements of aerosol particulates could be considered as a baseline value. Even here particles were found, including:

■ gypsum particles containing trace manganese or zinc – possibly from the metallurgy industry;

- fly-ash particles from power plants;

- iron and silicon rich particles with high zinc concentrations – probably produced during iron-manganese furnace processes; and

- titanium-rich particles with very high copper, zinc and zirconium concentrations – probably from coal-fired boilers, asphalt production and power plants.

Radioisotope XRF spectrometers

The X-ray excitation source in this instrument is radioactive (Figure 15). The fact that no specialised component is needed for the primary source of radiation makes this instrument very compact and portable. It can be used in the field, for example, in determining the calcium content in cement clinker, assessing lake sediments, and in identification during scrap metal sorting. On-site sorting of leaded and unleaded glazed pottery on archaeological excavations is done using radioisotope XRF.

The intensity of a radioactive source falls over time and this constrains the accuracy, precision, detection limits and the number of elements that can be detected. Occasional re-calibration is needed.

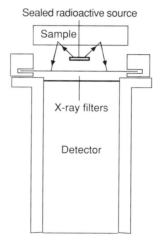

Figure 15 A sectoral view of a radioisotope X-ray probe

After excitation the secondary radiation from the sample is separated by filters – *eg* nickel and copper filters transmit only K_{α} radiation from zinc. Radioactive sources can be changed for particular ranges of elements. Plutonium-238 is the best radioactive source for the first row transition metals.

RS•C

Applications

Industrial

X-ray fluorescence has widespread industrial applications, particularly in the quarrying industry. Wavelength dispersive XRF is used for quality control in the limestone industry, monitoring the purity of calcium carbonate, calcium oxide and slaked lime. Samples are prepared as pressed powder pellets for qualitative analysis to give assurance that the chemicals are free from unwanted impurities. The material is then prepared as a fused bead for quantitative analysis. Similar monitoring, where the elemental composition must conform to strict specifications, is carried out for magnetic alloys processed for use in meters, motors, generators, actuators and switches.

Archaeological artifacts

Energy dispersive XRF is widely used for analysing antiques. The X-ray beam is collimated so that areas under 2 mm^2 can be analysed. X-ray fluorescence analysis, as a surface technique, can be performed without physically disturbing the sample, but in many cases the surface of an artifact has to be prepared by removing deposits such as rust and chemicals leached from the interior of the sample. However, this can be advantageous when identifying surface components such as platings.

Fire gilding

Fire gilding is an ancient method of applying gold amalgam – an alloy of gold and mercury – to a metal surface. The amalgam is heated to evaporate excess mercury leaving a layer of gold. Analysis by EDXRF has shown residual traces of mercury in Roman artifacts, which indicates that fire gilding was used by the Romans.

Pigments

Pigments are used to decorate ceramics. Historical changes in the production and supply of pigments can be traced by identifying differences in composition using qualitative EDXRF. Cobalt blue is the pigment on the underglaze of pre-15th century blue and white Chinese porcelain. Cobalt is scarce in China, which suggests that the raw materials for the pigment were imported. Later artifacts are associated with Chinese sources of manganese, such as in the mineral asbolane, $(CoMn)O.2MnO_2.4H_2O$. These analyses correlate with documentary evidence that the Chinese imported cobalt from Persia before the 15th century, after which they manufactured porcelain pigments from their own manganese-based sources.

Question

An environmental estimation was done on the proportion of lead in roadside plants. It is known that 500 ppm lead is lethal to plants like nettles. Samples of nettles were collected and ground down. Known amounts of lead powder were added to each sample, mixed and analysed using XRF with the instrument set at the Pb K_α peak. The following data were obtained:

% added lead powder	Counts/10s
0.2	7301
0.6	8997
1.0	10684
1.4	12410
1.8	13989

1. What is the lead content of the nettles in ppm?

2. How could you account for the results in terms of the lead toxicity?

RS•C

Answer

1. A standard addition graph (Figure 16) is drawn of percentage added lead against counts per 10s. Extrapolating the graph gives the lead content as 1.5% – *ie* 1.5 parts per hundred. In terms of ppm this translates to 15 000 ppm, thirty times greater than the lethal level.

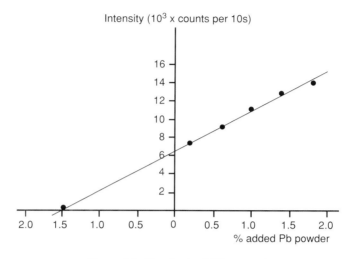

Figure 16 Standard addition graph

2. Since the samples were collected from live nettle specimens, most of the lead cannot have been absorbed by the plants – *ie* it was on the surface of the plant. The most likely source would be vehicle emissions. To determine the lead absorbed, the samples would have to be completely washed.

RS•C

This page has been intentionally left blank.

RS•C

Elemental microanalysis

Introduction

Elemental microanalysis is a technique for finding the relative amounts of carbon, hydrogen, nitrogen, sulfur and oxygen in both organic and inorganic compounds. There are five main stages in the analysis:

■ weighing;

■ combustion to form gaseous oxides (the products are converted to single element oxides with the exception of the nitrogen oxides which are then reduced to the elemental form);

■ separating the gaseous oxides and nitrogen;

■ detecting the gases; and

■ quantifying the results.

The traditional method for determining the proportion, by mass, of elements in organic compounds is to burn carefully weighed samples of the compounds and pass the gases through absorption tubes. The tubes are checked for gain in mass, and the amounts of carbon, hydrogen, nitrogen and sulfur are determined. The mass of oxygen is calculated by subtracting the masses of the other elements from the total mass of the sample. These gravimetric methods are laborious, time-consuming and relatively inaccurate (ca ± 10%) compared with modern automated techniques which have an accuracy of ca ±0.1%.

In modern elemental analysers, times for analysing, from sampling to data print-out, vary from three to 15 minutes depending on the instrument and the mass and type of sample. Instruments can be configured to include :

■ the determination of oxygen; and

■ the determination of nitrogen alone.

Other constituents of the sample, such as halogens and water, are determined using other methods.

Consider a pharmaceutical chemist synthesising a medicine for lowering blood pressure. The first step is to confirm that all the elements are present in the expected proportions corresponding to the molecular formula. A known mass of the product is analysed and the proportions of each element present are recorded. If the result is different from that expected, another compound may need to be made or the original sample is checked for contamination. If the result corresponds with the predictions, then the follow-up strategy is to confirm the structure using other techniques such as infrared (IR) spectroscopy, mass spectrometry (MS) and nuclear magnetic resonance (NMR) spectroscopy.

Coupling the elemental analyser to a mass spectrometer extends the analytical possibilities by providing isotopic ratios of the elements, in addition to the information about the amounts of each element.

RS•C

Instrumentation

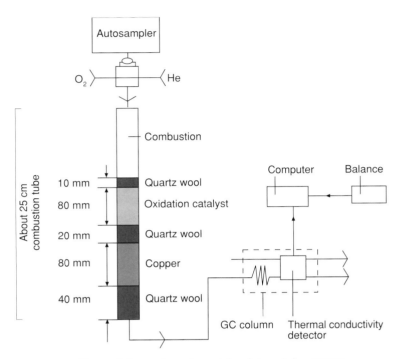

Figure 1 An analyser layout for determining CHNS

There are three main components in the elemental analyser (Figure 1):

■ the combustion tube;

■ the gas chromatography (GC) column; and

■ the detectors.

Between the autosampler and the combustion tube (Figure 2) there are electronically controlled inlets for the carrier gas – normally helium – and the oxygen. A typical combustion tube is made of a heat-resistant material such as quartz or silica. This combustion tube is normally *ca* 25 cm long and 2 cm in diameter and is operated at about 1000 °C. The upper part of the combustion tube is packed with an oxidation catalyst such as chromium(IV) oxide. The advantages of this catalyst are that:

■ it is refractory (heat resistant);

■ it does not adsorb the gases; and

■ it does not catalyse the oxidation of nitrogen.

Pieces of copper wire packed into the lower end of the combustion tube reduce the oxides of nitrogen and remove any excess oxygen.

$$2Cu(s) + 2NO(g) \rightarrow 2CuO(s) + N_2(g)$$

$$2Cu(s) + O_2(g) \rightarrow 2CuO(s)$$

RS•C

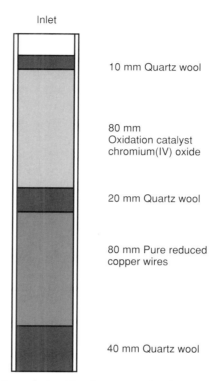

Figure 2 Reactor packing for CHNS simultaneous determination

The combustion tube usually has to be changed after *ca* 250 samples have been analysed.

The gases pass out of the combustion tube into narrow tubing, about 3 m long with an internal diameter of 1 mm, then through a GC column where they are separated, then detected by thermal conductivity. Some analysers use infrared detection. Some gases absorb infrared radiation at precise wavelengths, and the energy changes are detected and quantified. The detector and the balance for weighing the samples are interfaced to a computer to display the data. An outlet tube transports the gases for disposal. Alternatively they pass into a mass spectrometer for further analysis.

RS•C

Principle/theory

The instrument is calibrated and recalibrated during analysis using a standard – *eg* sulfanilamide (4-$(H_2N)C_6H_4SO_2NH_2$). Sulfanilamide contains carbon, hydrogen, nitrogen and sulfur in the following proportions by mass:

C = 41.84%
H = 4.68%
N = 16.27%
S = 18.62%

The standard should have as close a composition as possible to the sample. Phenanthrene ($C_{14}H_{10}$), is used for samples with a low nitrogen and sulfur content and urea (H_2NCONH_2) (proportion of carbon = 20%) is used for compounds with a low carbon content. Inorganic compounds pose a specific problem because it is difficult to find inorganic standards, such as calcium carbonate ($CaCO_3$), which burn completely. Recalibration occurs at frequent intervals, usually after every five samples. The detection limits for each of the elements is as low as 10 parts per million (10 ppm or 10^{-5} g g^{-1}).

Between 0.1 mg and 100 mg samples are weighed in a small metal capsule which is then crimped and placed in the carousel of the autosampler. Accurately weighed masses are entered into the computer database. The capsules are usually made of silver, platinum or tin. Tin has two advantages over the other two metals: it is cheaper and plays an important part in the combustion process (Box 1). Crimping the capsule reduces the amount of air present because excess nitrogen could affect the result. Volatile liquids such as fuels are hermetically sealed (made air-tight by melting the glass) into glass vials by a device supplied by the manufacturer.

As the sample drops into the combustion chamber it is purged with a flow of helium, which acts as a carrier gas and further displaces elemental nitrogen .

Combustion of the sample occurs once oxygen is injected into the carrier gas. The temperature of the combustion tube is maintained at 950–1030 °C depending on the instrument and the sample being analysed. On entering the combustion tube the sample is instantaneously atomised, the atoms combining with oxygen in quantities, the ratios of which reflect the stoichiometric composition of the compound.

Tin capsules   **Box 1**

The use of tin as the metal for the container capsule helps to maintain the high temperature of the exothermic reaction necessary for combustion to occur.

The oxidation of tin can be described as:

$$Sn(s) + O_2(g) \rightarrow SnO_2(s) \qquad \Delta H_c = -142 \text{ kJ mol}^{-1}$$

The enthalpy of combustion of tin contributes to the overall reaction by raising the temperature to over 1800 °C. In addition, the product, tin(IV) oxide, is an oxidation catalyst, thereby increasing the overall reaction rate. The high temperatures generated by combustion means that a wide variety of materials can be analysed, such as blood, plastics, foods, oils and pharmaceuticals. Materials such as rocks and alloys may need special preparation. Rocks are particularly problematic because they have high melting points and samples do not oxidise easily. It is therefore essential to digest the sample first and to adjust the instrument for the correct combination of catalysts, gas flow rates and furnace temperatures. Even then the problems are considerable because the combined amounts of carbon, nitrogen and sulfur is often less than 1% by mass of the total sample.

RS•C

After the initial combustion in the combustion tube, the gaseous mixture passes downstream through a layer of catalyst to ensure complete conversion to its oxides. At the lower end of the combustion tube, the mixture passes through heated copper to remove any excess oxygen and to reduce nitrogen oxides to molecular nitrogen. Nitrogen oxides interfere with the detectors, hence nitrogen is formed instead.

Next, the gases are separated by passing them through a gas chromatography (GC) column. The eluted gases are detected by making use of the fact that they exert a cooling effect on a hot filament depending on their thermal conductivity. Four filaments are arranged in a Wheatstone bridge circuit (Figure 3). The helium carrier gas alone flows across two of the filaments, while the carrier gas plus the sample gas flows across the other two. This causes an imbalance in the bridge which is recorded as a voltage signal. (*Modern chemical techniques*, London: Royal Society of Chemistry, 1992, p.124)

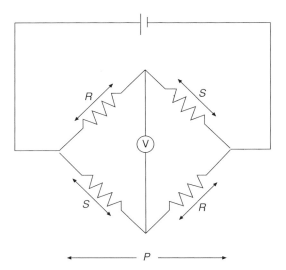

P = potential across whole bridge
R = potential across one filament in *reference* stream
S = potential across one filament in *sample* stream

Figure 3 A Wheatstone bridge arrangement of circuits

Carrying out an analysis

For routine analyses in modern analysers, the sample is weighed, placed in a capsule and inserted into the autosampler. Once the mass has been recorded and stored the rest of the analysis proceeds automatically. An adjustment has to be made for some analyses. An example is searching for trace amounts of hydrocarbons in sands for pollution monitoring. Sands with a low organic content do not burn readily. The catalyst used for predominantly inorganic samples is vanadium(V) oxide (V_2O_5). Trial-and-error adjustments of catalyst mass to sample mass are checked to find the best conditions for precise and accurate determinations. The temperature in the combustion tube has to be raised as high as possible but melting the copper (melting point 1083 °C) in the lower part of the tube has to be avoided.

The gas flow rate is altered to optimise the read-out from the detector. The peaks on the read-out corresponding to carbon and nitrogen are close together, hence a slow flow rate is used to maximise separation between the peaks (Figure 4).

RS•C

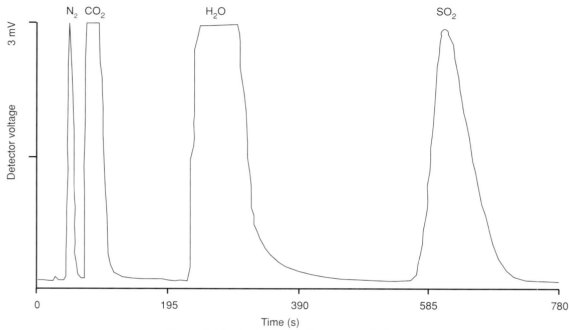

Figure 4 A typical printout from an analysis

Determining sulfur

Sulfur(IV) oxide (SO_2), is detected in elemental microanalysis, but further oxidation to sulfur(VI) oxide (SO_3) has to be prevented. Fortunately, the very high temperatures produced by the combustion process energetically favour the formation of SO_2. To ensure as high a yield of SO_2 as possible, the gases can be passed over a tungsten(VI) oxide catalyst. (Tungsten(VI) oxide has acidic properties so does not react with the sulfur oxides.) This is followed by passage over copper, which at high temperatures favours the reduction of sulfur(VI) oxide to sulfur(IV) oxide.

Determining nitrogen

Determining the total nitrogen content is important in food production, agriculture and the brewing industry, because it is an indicator of nutritional value in terms of protein. Analysers can be modified to determine nitrogen only. In modern analysers the total run time is 3–5 minutes and the mass of samples can be greater than for multi-element analysis – *ie* up to 0.5 g. The following features are characteristic of a dedicated nitrogen analyser.

■ Soda lime traps are incorporated into the reactor sequence to remove water, carbon dioxide and the sulfur oxides.

■ Larger volume flows of oxygen are needed for combusting larger amounts of sample.

■ There are lower detection limits for nitrogen.

■ Methionine ($CH_3SCH_2CH_2CH(NH_2)CO_2H$) or aspartic acid ($HO_2CCH_2CH(NH_2)CO_2H$) is used as the calibrant.

■ With fatty substances, methane is sometimes produced as a result of the high mass of carbon to mass of hydrogen ratio. However, methane can be oxidised and the products removed.

RS•C

An alternative method of detecting nitrogen is chemiluminescence, which detects trace levels of nitrogen in parts per million (ppm) quantities. The nitrogen is oxidised to nitrogen monoxide (NO), then oxidised further by ozone to nitrogen dioxide (NO_2). This is formed in an excited state. The excited molecules revert to their ground state and are quantified spectroscopically.

The Kjeldahl method   **Box 2**

The traditional laboratory method for accurately determining nitrogen in proteins and other nitrogen-containing compounds is the Kjeldahl method. In this process the material is digested by sulfuric acid in the presence of a catalyst, usually copper, and potassium sulfate (the latter to increase the boiling point and speed up the digestion further). An example of the reaction is the digestion of 4-aminobutanoic acid:

$$NH_2(CH_2)_3CO_2H + 10H_2SO_4 \rightarrow 4CO_2 + 9SO_2 + 12H_2O + NH_4HSO_4$$

The solution is then cooled. Ammonia is distilled off by the addition of concentrated alkali and absorbed into a standard solution of excess acid. The acid is back-titrated with a standard base and the amount of nitrogen is calculated.

Modern nitrogen analysers have considerable advantages over the Kjeldahl method. They are quicker – a Kjeldahl analysis can take at least one and a half hours and, since they are automated, they are less costly on technician time. The Kjeldahl method also requires a fume cupboard and the disposal of toxic waste. Kjeldahl analysis is still performed in laboratories where nitrogen analysis is not routine, because it is as accurate as a modern analyser and does not involve the expense of buying an analyser.

The Kjeldahl method is often better than an elemental analyser for determining inorganic nitrogen. For example, for low amounts – less than 0.3% m/m – in ammonium (NH_4^+) deposits in the mineral buddingtonite.

Determining oxygen

Organic oxygen is converted to carbon monoxide in the elemental analyser, then separated from other gases using gas chromatography (GC) and detected by thermal conductivity. The sample is passed through a high temperature pyrolysis furnace containing nickel-coated carbon heated at 1060 °C. Decomposition occurs, and the oxygen that is released combines with the carbon to form carbon monoxide, which is swept through the analyser by a stream of helium and converted to carbon dioxide before being measured.

During pyrolysis a number of other gases may be formed in addition to carbon monoxide. Any acidic oxides are absorbed by passing through an alkaline scrubber, then the remaining gases are flushed through a GC column where they are separated and individually detected by thermal conductivity.

For trace analysis of oxygen content in organic compounds, the carrier gas is doped with halocarbon vapour, which accelerates their decomposition.

The level of oxygen in fuels, for example the levels of oxygenates (Box 3) in lead-free petrols, often need to be determined. In motor racing, the proportion of oxygen in racing car fuels is checked, and very heavy fines are handed out for exceeding strict limits.

RS•C

Oxygenates   **Box 3**

Oxygenates are organic oxygen-containing compounds such as alcohols and ethers. Branched chain oxygenates are added to car fuels to increase their octane rating – *ie* to improve their anti-knock performance. One of the main oxygenates used is methyl tertiary butyl ether (MTBE) (2-methoxy-2-methylpropane).

Coupling to a mass spectrometer

Elemental analysers can be interfaced with a mass spectrometer. When investigating the composition of a soil, elements such as isotopically labelled nitrogen (^{15}N) can be tracked over a period of time in various crops. In the perfume industry, perfumes can be authenticated both by checking their characteristic elemental composition and by comparing isotopic ratios such as ^{12}C:^{13}C. Some raw materials come from peppermint and spearmint plantations in the US and lavender fields in the south of France. The raw materials for products like perfumes have characteristic isotopic fingerprints depending on the types of plants from which the ingredients are extracted and their location.

Applications of elemental analysis

Food

Manufactured dairy products must not degrade quickly and lose nitrogen as protein. The protein content of synthetic and natural dairy products can be determined and followed by using nitrogen analysis. For example, consider coconut milk that is manufactured for the Indonesian market. Its formulation, taste and nutritional content must match that of milk extracted locally from fresh coconuts. The advantage of the manufactured product is that it is more durable. Analysis can provide accurate data for quality control to assess that the protein content is similar in both the synthetic and the natural milk.

In the flour milling and animal feed industry, maintaining high amounts of protein in products is paramount because the price of wheat is related to protein content.

Agriculture

Elemental analysers can be used to assess the effects of crop rotation and prolonged use of soil on the organic carbon content of the soil. By interfacing an analyser with a mass spectrometer the movement of carbon – in the form of ^{13}C – in the soil can be followed by sampling over a period of time. Nitrogen is an essential nutrient for plant life and animal life. To maintain the flow of nitrogen in the soil, fertilisers must have the right composition to match the changing nitrogen content. A fertiliser with the wrong composition results in reduced crop yield. Elemental analysis has a role in quality control in evaluating fertilisers and composts for C, N and S content, and matching the performance of nitrogen in the soil. Since raised levels of sulfur in soil can indicate acid rain pollution, elemental analysis is a sensitive technique for detecting this. Geological sediments can also be characterised by elemental analysis and related to crop yields and food chains.

Other uses

Elemental analysis is important in the plastics industry and for carbon fibres, ceramics, glass-fibres and semiconductors. In the metals industry the proportion of carbon in steels also needs careful monitoring.

RS•C

Oil field prospecting relies on many different techniques. However, reliable indications about the whereabouts of viable oil fields are given by the proportion of carbon in samples, and by the $^{12}C : ^{13}C$ ratios. An analyser interfaced with a mass spectrometer is ideal for guiding initial prospecting.

Analysing other elements

Organic materials often contain chlorine, bromine, iodine and fluorine as well as CHNS/O. These elements can be determined by combustion in an oxygen flask (Figure 5) followed by absorption in an alkali and titration.

Combustion flasks

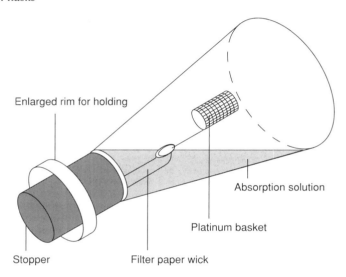

Figure 5 An inverted combustion flask

An absorption solution, usually sodium hydroxide or sodium chlorate(I), is placed in the flask before combustion. The flask is then filled with oxygen gas to displace the air. The sample is placed on the filter paper and is placed in the platinum basket – which acts as a catalyst for the reaction. When the filter paper is lit, the stopper is inserted and held firmly closed by gripping the joint. The flask is gently inverted allowing the absorption solution to close the joint.

After the combustion, which should take about 30 seconds, the filter paper and sample should be completely burned. Once the flask has cooled, it is gently shaken while still inverted. The flask can then be opened and the element determined by titration. The processes are summarised in Table 1.

RS•C

Reagent	Chlorine	Bromine	Iodine	Fluorine
Absorption solution	Sodium hydroxide $(0.05 \text{ mol dm}^{-3})$	Sodium chlorate(I) (1 mol dm^{-3})	Sodium hydroxide $(0.3 \text{ mol dm}^{-3})$	Sodium hydroxide $(0.05 \text{ mol dm}^{-3})$
Solvent	Propan-2-ol	Water	Water	Water
Titrant	Mercury(II) nitrate, $Hg(NO_3)_2$ $(0.005 \text{ mol dm}^{-3})$ and nitric acid $(0.1 \text{ mol dm}^{-3})$	Sodium thiosulfate $(0.05 \text{ mol dm}^{-3})$	Sodium thiosulfate $(0.05 \text{ mol dm}^{-3})$	Lanthanum(III) nitrate
Indicators	Bromophenol blue (BB) Diphenylcarbazone (DPC)	Iodine/Starch	Starch	Potentiometric titration
Colour changes	BB – blue to yellow; DPC – pale yellow to faint violet	Blue to colourless	Blue to colourless	
Other reagents	Barium nitrate for removal of sulfur in sulfur-containing compounds	Ammonium molybdate (V) as catalyst; potassium iodide	Bromine to convert iodine to iodate(V); methanoic acid to remove excess bromine; methyl red to confirm removal of excess bromine	
Any special conditions				A quartz flask is used rather than borosilicate glass which reacts with fluorine.
Equations		$BrO_3^- + 6I^- + 6H^+$ $\rightarrow 3I_2 + 3H_2O + Br^-$	1. $I_2 + 5Br_2 + 6H_2O$ $\rightarrow 2IO_3^- + 10Br^- + 12H^+$ 2. $HCOOH + Br_2$ $\rightarrow 2HBr + CO_2$ 3. $IO_3^- + 5I^- + 6H^+$ $\rightarrow 3I_2 + 3H_2O$	

Table 1 A summary of titration processes for halogens

RS•C

Determining water content

The amount of water in an organic material needs to be known for a variety of reasons including:

- determining how much water is in foodstuffs;

- finding out how much water there is in organic solvents;

- determining the functional groups in reactions which lead to the quantitative production of water; and

- the percentage by mass of water in organic hydrates.

The standard method for determining water content in a sample is known as the Karl Fischer (KF) titration. In this technique a salt is formed from an organic amine and is oxidised stoichiometrically by iodine in the presence of water. Traditionally the reaction is based on pyridine, but more recently commercial reagents have come on the market which contain relatively harmless odourless amines. The reaction is usually done in methanol solution. The reaction occurs in two stages:

1. $CH_3OH + SO_2 + RN \rightarrow RNH^+ CH_3 SO_3^-$

2. $H_2O + I_2 + RNH^+ CH_3SO_3^- + 2RN \rightarrow RNH^+ CH_3SO_4^- + 2(RNH^+)I^-$

The redox changes are:

$CH_3SO_3^- + H_2O \rightarrow CH_3SO_4^- + 2H^+ + 2e$

$I_2 + 2e \rightarrow 2I^-$

The end-point is detected either by a colour change or by a change of conductivity using an auto-titrator. A yellow $SO_2–I_2$ complex is formed at pH 6–7 which decomposes bringing about decolorisation.

The burette is filled with the titrant and the methanolic solution is added to the titration flask. An initial titration is done to remove any water that may be present as an impurity. The sample is then added to the titration flask and the water content is determined.

The stability of the titrant is checked regularly by using a standard. This is a precaution against the titrant absorbing atmospheric moisture. A good standard is sodium tartrate dihydrate, $(NaO_2CCH(OH)CH(OH)CO_2Na.2H_2O)$ which is stable at room temperature and atmospheric pressure, does not effloresce (become covered with a powdery crust), and is not hygroscopic.

The system has to be adjusted to maintain the accuracy of the titration. A low pH inhibits the reaction, whereas an alkaline medium produces side reactions. A pH of 4–7 is maintained in the auto-titration using an appropriate buffer.

RS•C

This page has been intentionally left blank.

RS•C

Gel electrophoresis

Introduction

The term gel electrophoresis covers a range of techniques that is used to separate, analyse and purify mixtures of biological molecules such as proteins and nucleic acids. These techniques can be adapted to:

- measure the relative masses of macromolecules;

- prepare nucleic acids and polypeptides for sequencing the component monomers – *eg* purine and pyrimidine bases, amino acids; and

- separate proteins, so antibodies can be raised.

The sample mixture is placed in a gel and is separated by applying an electric field to the gel, which is soaked in a liquid buffer (Figure 1). The gel is a sponge-like structure based on a three-dimensional polymeric network. It has the texture of a jelly. Electrophoresis can be done in a variety of media including liquids and soaked paper, but the properties of gels can be controlled during their preparation, and they are more chemically stable as a support medium.

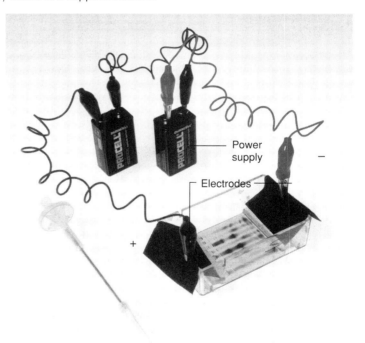

Figure 1 Basic arrangement for electrophoresis

The components in the sample mixture are charged because proteins carry either an overall positive or an overall negative charge, depending on the pH of their environment. Nucleic acids such as deoxyribonucleic acid (DNA) are usually negatively charged at the pH used for their separation using electrophoresis. The molecules move in response to an electrical field applied across the mixture. The rate of progress of the molecules depends on their size, charge and shape. On separation,

RS•C

the components are concentrated into bands or zones and appear to move along straight tracks during separation.

Each band or zone can be quantified using a variety of methods. Using sensitive techniques, such as bioassays or silver-staining, amounts as small as 10^{-18} g can be detected.

Gel electrophoresis has a variety of applications such as checking the adulteration of foods, chromosome sequencing, DNA fingerprinting and characterising the chemicals responsible for allergic symptoms. Gel electrophoresis is used to screen infants' milk for α-lactoglobulin, a protein which is lethal for small babies.

Principles and instrumentation

The main components used in gel electrophoresis are:

■   an electrophoresis chamber;

■   a gel support medium soaked in conducting buffer;

■   a means of generating an electric field – *eg* a power pack;

■   probes for detecting and/or measuring the separated molecules; and, if necessary,

■   a means of extracting the individual products.

A very basic electrophoretic separation system consists of a gel medium in buffer linked by a contact bridge to electrodes (Figure 2).

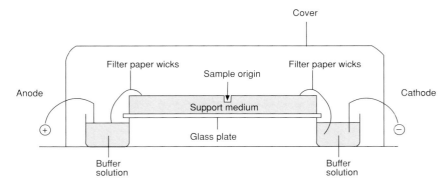

Figure 2 Basic horizontal gel electrophoresis system

The mixtures are placed in small wells in the gel medium and when the electric field is applied the components of the mixture separate out. Their rate of migration across the gel becomes constant when the force of attraction between the electrode and the oppositely-charged component is equal to the frictional force of the gel medium resisting the motion for that particular species. The relative rates at which components move and therefore the extent to which they separate are influenced by the strength of the field, the nature of the gel, and the surrounding buffer. Each species responds to these factors according to what is called its electrophoretic mobility (μ). This is a constant for each species.

As a component migrates through an electric field its rate of migration, or velocity, depends on the strength of the electric field. As the field strength is increased, by raising the voltage across the gel, the velocity of the migrating species increases proportionately.

RS•C

Thus $v \propto E$

where
v is the migration velocity of a particular species in cm s^{-1};
and E is the applied field strength in V cm^{-1}

Or, $v = \mu E$

where
μ is the electrophoretic mobility for a particular species in cm^2 V^{-1} s^{-1}.

The two main separation parameters are size and charge. Smaller molecules have higher electrophoretic mobilities and migrate more quickly than larger molecules (Figure 3a) and for a particular size of molecule, those with a higher charge move faster (Figure 3b).

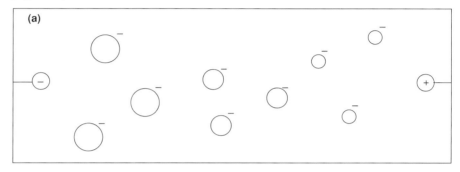

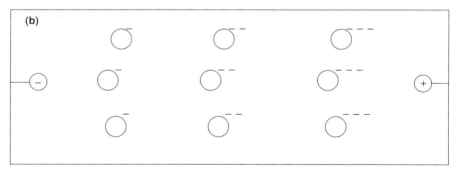

Figure 3 (a) Smaller molecules move faster in an electric field than larger molecules with the same charge (b) Molecules of the same size but with a higher charge move faster

Gel preparation

In electrophoresis, the gels are prepared in slabs or rods, so they can be used for horizontal or vertical separation. Rods are usually cylinders of gel kept in glass tubes (5 mm diameter) (Figure 4), whereas slabs are between 0.75 mm and 1.5 mm thick, and are usually rectangular in shape.

RS•C

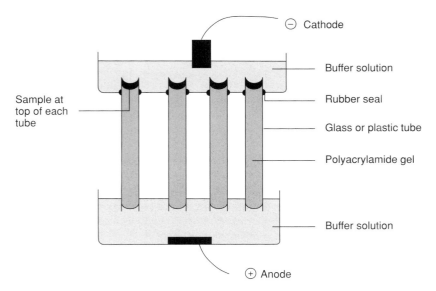

Figure 4 Apparatus for tube gel electrophoresis

Factors affecting mobility   **Box 1**

Voltage. The velocity of a molecule is directly proportional to the voltage gradient across the gel.

Size. Smaller molecules migrate more quickly than larger molecules carrying the same charge.

Charge. For a particular size of molecule, those carrying a higher charge move faster.

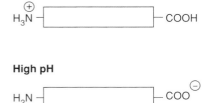

Figure 5 The ionisation of amino and carboxyl groups at low and high pH

Buffer. Proteins can exist as zwitterions that are either positively or negatively charged because they contain both acidic and basic groups. The ionisation of the protein depends on the pH of the buffer (Figure 5).

In electrophoresis, the buffering action (pH) must be kept constant to maintain the charge on the protein and to keep the mobilities constant.

Shape. The shape of a molecule influences its migration through the gel. A molecule with a lot of side chains experiences more frictional resistance than a linear molecule of the same mass and charge, and will therefore move more slowly.

Temperature. Temperature control is critical during electrophoresis. A rise in temperature can denature the proteins – *ie* change the nature of their properties. Uneven heat distribution in the gel distorts band shapes due to different mobilities at different temperatures. The temperature is controlled by water cooling or by running the electrophoresis at low voltages.

Power. Power supplies for electrophoretic separations hold either the power, voltage or current constant. Any change in these parameters generates extra heat during a run. Conventionally, the separation of proteins is carried out at constant current, nucleic acid separations at constant voltage, and DNA fingerprint gels at constant power.

$P = IV$ or $P = I^2R$ or $P = V^2/R$

where
P = the power measured in watts V = the voltage measured in volts
I = the current measured in amperes R = the resistance measured in ohms.

RS•C

Slab gels are the most common because their shape makes it easy to separate many samples at the same time. Because any rise in temperature can disturb successful separation, the relatively large surface area of a slab gel is an advantage because it dissipates any generated heat. However, rod gels can be cut transversely into very thin slices, and are useful for analysing components using radioactive labels.

Polyacrylamide and agarose gels are the most commonly used gels in research and industrial laboratories.

Polyacrylamide gels
The main advantage of polyacrylamide gels is that their preparation can be controlled closely to produce pores of a particular diameter. These pores have a sieving effect (Figure 6) and provide frictional resistance to the migration of large molecules.

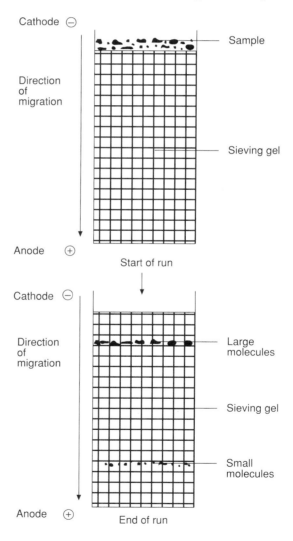

Figure 6 A schematic diagram to illustrate a sieving effect

The cross-section of the pores can be controlled by modifying the concentration of the gel. Generally, polyacrylamide gels have smaller pore sizes than agarose gels, but the two gels can be combined to provide a high resolution separation for a wide range of molecular sizes.

RS•C

Preparing polyacrylamide gels Box 2

Preparing polyacrylamide gels involves monomers, crosslinkers, initiators and radicals. The chemicals and their functions are described in Table 1.

Chemical	Formula	Function
Acrylamide (propeneamide)	$CH_2CHCONH_2$	Monomer – the repeating unit of the polymer
N,N'-methylenebisacrylamide	$CH_2CHCONHCH_2NHCOCHCH_2$	Crosslinking agent
Ammonium peroxodisulfate (ammonium persulfate)	$(NH_4)_2S_2O_8$	Initiator - generates free radicals for the polymerisation process
Riboflavin (vitamin B2)	[not shown]	Initiator - the generation of free radicals is initiated by light
N,N,N',N'-Tetramethylene Diamine (TEMED) (N,N,N',N'-tetramethyl-1,2-diaminoethane	$(CH_3)_2NCH_2CH_2N(CH_3)_2$	Catalyst for the formation of radicals from ammonium peroxodisulfate

Table 1 Chemicals used in the synthesis of polyacrylamide

Polyacrylamide (polypropenamide) is made by polymerising acrylamide (propenamide, $CH_2CHCONH_2$) monomers with the crosslinking agent N,N'-methylenebisacrylamide, ($CH_2CHCONHCH_2NHCOCHCH_2$), known as bisacrylamide (Figure 7).

Figure 7 The polymerisation of acrylamide

The crosslinking agent is also instrumental in determining the mechanical properties of gels, including pore size. Without the crosslinking agent, the polymerised gel would be a random array of fibres. Two parameters are used to indicate the effective pore size of the gel.

■ The total concentrations (T) of monomers (mass per unit volume).
■ The percentage by mass of the crosslinking (C) monomer (for polyacrylamide, bisacrylamide).

These two parameters are known respectively as %T and %C, respectively.

Increasing the concentration of either acrylamide or bisacrylamide decreases the effective pore size in the polyacrylamide gel. For the molecular sieving of large molecules, such as ribonucleotides, which have relative molecular masses of about 10^6 atomic mass units (amu), a low concentration of acrylamide is used. This results in a rather liquid gel which can be made more viscous by adding up to 0.5% by mass of agarose.

The pore size in polyacrylamide gels is also reduced by increasing %C – diameters as low as 20 nm can be attained. The %C value has an upper limit of about 30% because a value exceeding this leads to gel turbidity and syneresis – a state in which the gel contracts and starts to exude liquid!

Polyacrylamide gels are commonly used for the fractionating of proteins, sequencing and fingerprinting DNA but, unlike agarose, they need particularly careful handling. The monomers can be absorbed through the skin and are linked to cumulative nerve damage.

RS•C

Agarose gels

Preparing agarose gels from Agar – a seaweed extract – is much simpler and less toxic than preparing polyacrylamide gels. Agarose is a polysaccharide (Figure 8), with the links between the polymer chains arising from hydrogen and hydrophobic bonds (Figure 9a).

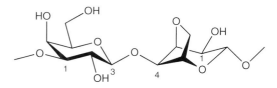

Figure 8 The chemical structure of agarose

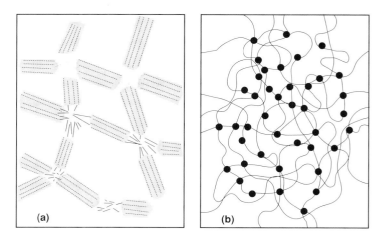

Figure 9 Agarose and acrylamide gels (a) Agarose gels form by hydrogen and hydrophobic bonds between long sugar polymers (b) Acrylamide gels have covalent crosslinks between polymer strands

Agarose gels are made by dissolving agarose powder in a boiling buffer solution. This solution is allowed to cool, and sets into a gel when the temperature drops to *ca* 40 °C. The higher the concentration of agarose in the gel, the smaller the pore size. Agarose concentrations range between 0.4% m/v to 4% m/v.

Techniques for separating proteins

There are two techniques commonly used for separating proteins. These are sodium dodecyl sulfate – polyacrylamide gel electrophoresis (SDS–PAGE) and isoelectric focusing (IEF).

The choice of a particular mode of separation and detection method depends on:

- the purpose of the separation;

- the chemical nature and size of the sample;

- the accuracy and precision needed for the analysis;

- time availability; and

- costs.

RS•C

The choice of which separation technique to use is a problem-solving exercise relying on some trial and error, exploratory runs and the experience of the practitioner.   For example, where a quick separation is needed it may be better to use a high field potential (voltage). On the other hand, this generates heat faster, which may spoil the separation. If the cooling system can cope with the extra heat then using a higher field potential is feasible.

Sodium dodecyl sulfate – polyacrylamide gel electrophoresis (SDS–PAGE)

This system involves dissociating proteins into their polypeptide subunits using an ionic detergent called sodium dodecyl sulfate ($CH_3(CH_2)_{10}CH_2OSO_3^-$ Na^+, SDS), then running the proteins on a polyacrylamide gel.

The protein is heated at 100 °C with excess SDS and a thiol reagent ie the reagent contains an –SH functional group such as 2-mercaptoethanol, (2-hydroxyethylmercaptan, $HSCH_2CH_2OH$), which cleaves the sulfur-to-sulfur bridges in protein molecules.

The SDS binds to the polypeptide molecule effectively coating it in a negative charge. The charge density is the same for each molecule, hence the only difference between the polypeptides is size. During the run the smaller polypeptides migrate towards the anode at a faster rate than the larger molecules. This difference is enhanced by the sieving effect of the polyacrylamide gel.

This type of separation is advantageous because it is done using only one variable – ie molecular size. However, two molecules of near identical size may not necessarily be separated, and the shape of the molecule also influences the rate of migration.

Concentration gradient polyacrylamide gels   **Box 3**

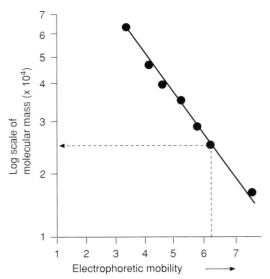

These gels have a gradient with an increasing acrylamide concentration along their length. The particular gradient of the gel can be adjusted, depending on the type of proteins being analysed. There are two main advantages to a concentration gradient gel. First, it gives sharper banding patterns. As the proteins migrate they come across smaller and smaller pore sizes. For a certain size of protein, the pore size reduces the rate of migration such that the advancing edge of the migrating band is slowed down more than the trailing edge. Secondly, it increases the range of molecular masses that can be measured on one gel, because the range of pore sizes allows for greater fractionation. Molecular mass markers can be used and a graph can be drawn for calibration purposes (eg Figure 10).

Figure 10 The estimation of molecular mass by SDS–PAGE. The line constructed through the black points corresponds to the mobilities of proteins of known molecular mass. Plotting the mobility of a protein of unknown size yields a molecular mass of 25,000.

The molecular masses of proteins or polypeptides can be calculated using the pore size of the polyacrylamide gel used for separating the proteins. This is done by running marker proteins of known molecular mass on the gel (Figure 11) and plotting a curve of relative mobility against molecular masses.

Polypeptide	Molecular mass
Myosin (rabbit muscle heavy chain)	212,000
RNA polymerase (E. coli) β′-subunit	165,000
β-subunit	155,000
β-Galactosidase (E. coli)	130,000
Phosphorylase a (rabbit muscle)	92,500
Bovine serum albumin	68,000
Catalase (bovine liver)	57,500
Pyruvate kinase (rabbit muscle)	57,200
Glutamate dehydrogenase (bovine liver)	53,000
Fumarase (pig liver)	48,500
Ovalbumin	43,500
Enolase (rabbit muscle)	42,000
Alcohol dehydrogenase (horse liver)	41,000
Aldolase (rabbit muscle)	40,000
RNA polymerase (E. coli) α-subunit	39,000
Glyceraldehyde-3-phosphate dehydrogenase (rabbit muscle)	36,000
Lactate dehydrogenase (pig heart)	36,000
Carbonic anhydrase	29,000
Chymotrypsinogen A	25,700
Trypsin inhibitor (soyabean)	20,100
Myoglobin (horse heart)	16,950
α-Lactalbumin (bovine milk)	14,400
Lysozyme (egg white)	14,300
Cytochrome C	11,700

Figure 11 Molecular mass markers

Continuous and discontinuous systems   **Box 4**

There are two types of buffer systems in an electrophoresis run – continuous and discontinuous. In a continuous system the same buffer, and hence the same pH, is used in both the gel and the electrode reservoirs. Discontinuous systems have different buffers, hence a different pH in the electrode reservoirs and the gel. There are also two gels – a stacking gel that is non-resolving and has large pores; and a resolving gel with small pores. The stacking gel is layered over the resolving gel.

The advantage of the discontinuous system is that it enables the loading of larger volumes of dilute protein samples while still maintaining good resolution.

The mechanism of a discontinuous system

In discontinuous systems, the proteins' mobility in the stacking gel is intermediate between the mobilities of two buffers. This has the effect of concentrating the proteins into a very narrow zone between them. When the proteins reach the resolving gel their mobility is reduced due to the sieving effect caused by the smaller pore size. The proteins unstack as they migrate through the resolving gel, separating into sharp bands according to their size.

RS•C

Instrumentation

Individual instrument design varies considerably, from inbuilt systems with injection-moulded electrophoresis tanks, to tanks and gel holders that can be constructed in the laboratory from accessible materials. However, all systems have similar components.

- Holders to keep the gel in place

- Tanks

- Electrodes to make contact with the gel

- Small plastic combs to make sample wells so that several samples can be run at the same time

- Micro-applicators to load the sample

- A power pack capable of supplying the necessary voltage and current.

Manufactured systems are likely to contain some kind of cooling system to dissipate the heat generated from an electrophoretic run. Most tanks come in a range of sizes. Some can hold up to four slab gels so that up to 200 samples can be run at any one time.

Running a sample in SDS–PAGE electrophoresis

Performing this type of electrophoresis involves:

- pouring the fully-prepared gel into the electrophoresis chamber before it sets;

- preparing the buffer;

- loading the samples on to the gel with the 'shark's tooth' comb (see p. 88); and

- running the electrophoresis at a constant voltage or current. (A constant current is normally used for separating proteins).

Pouring the gels. For a continuous system, once the ingredients for a particular concentration of polyacrylamide gel have been prepared and mixed they are poured between two clean glass plates, which form the gel mould.

In a discontinuous system, the resolving gel is poured first and allowed to polymerise – this takes up to 30 minutes. The stacking gel is poured after the resolving gel has polymerised. A small amount of the stacking gel mixture is run over the resolving gel. The comb is inserted into the stacking gel mixture before it polymerises, then it is removed to leave the sample wells. The sample wells are rinsed over with reservoir buffer, and then filled with buffer. The slab gel can be left overnight before use providing dehydration is avoided – *eg* filling with buffer and covering with cling film, or simply leaving in a refrigerator or cold room.

Sample preparation. The sample usually consists of a few drops of protein. This is mixed with 0.0625 mol dm^{-3} of Tris-HCl (buffer), at pH 6.8, added to 2% v/v sodium dodecyl sulfate (SDS), 5% v/v 2-mercaptoethanol, 10% v/v sucrose or glycerol and 0.002% v/v bromophenol blue as the tracking dye.

Substance	Concentration	Purpose
Tris-HCl at pH 6.8	$0.0625 \text{ mol dm}^{-3}$	Buffer. To maintain conductivity throughout the gel and to resist pH changes
Sodium dodecyl sulfate, SDS.	2% by volume	Splits the protein into polypeptide sub-units and binds to the polypeptide quantitatively. Each polypeptide has a constant negative charge density
2-Mercaptoethanol	5% by volume	Breaks the sulfur-to-sulfur bridges in proteins
Sucrose	10% by volume	Increases density of sample so that it sinks into the sample wells and does not mix with the cathode buffer
Propane 1,2,3 triol (glycerol)		Glycerol can be used as an alternative to sucrose
Bromophenol blue	0.002% by volume	Tracking dye. Shows the positions of the smallest components so electrophoresis can be terminated before any components run off the gel

Table 2 The functions of the various components of the mixture run in SDS–PAGE

Dithiothreitol ($C_4H_{10}O_2S_2$, DTT) is sometimes used instead of 2-mercaptoethanol because DTT is odourless and is not, unlike 2-mercaptoethanol, prone to auto-oxidation. It is important that the mole ratio of SDS to the polypeptide is at least 3:1 to ensure complete binding.

Just before electrophoresis, the mixture is heated in boiling water for 3 minutes to ensure the protein has unfolded completely. The mixture is then allowed to cool to room temperature. The presence of insoluble material, or reagents that have failed to dissolve, can lead to protein 'streaking' during electrophoresis. This mixture is centrifuged to remove any insoluble substances. The sample can then be used immediately or can be stored at -20 °C.

Running the sample. For SDS–PAGE – with the large negative charge of the detergent component – the anode is at the end opposite the sample wells. The electrodes are connected to the power pack and the electrophoresis is then run at a constant voltage or current. Compromise voltages and currents have to be used to avoid overheating on the one hand and slow separation on the other. Samples on slab gels can be left to run overnight in a water-cooled system, at room temperature.

SDS–PAGE can be carried out to detect the adulteration of cheese made from ewe's milk. This has important effects because some people suffer from allergic reactions to cow's milk. Standards are made up of cow's milk in ewe's milk and quantified by densitometry. Cheeses and yoghurts are often adulterated but there should not be more than 3% cow's milk present. A sample can be run in SDS–PAGE alongside the standards, and quantified.

RS•C

Isoelectric focusing (IEF)

The principle of this separation technique is the fractionation of molecules, predominantly proteins, according to their isoelectric points (pI values) (Box 5). Isoelectric focusing is a particularly useful technique for purifying and separating proteins and identifying enzymes. The technique is also used in many forensic and clinical applications.

Explaining pI   **Box 5**

The pI (where I = the **isolectric point**) of a zwitterionic molecule is the pH at which it has no net charge. Consider a protein, composed of a number of amino acids, which have a range of free carboxyl and amino groups. The amino groups become positively charged at low pHs, while the carboxylic acid groups remain unchanged, hence the protein is a positively charged ion. However, at high pH values the carboxylic acid groups become deprotonated and negatively charged while the amino groups are uncharged conferring an overall negative charge on the protein (Figure 12).

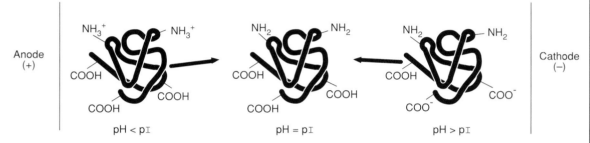

Figure 12 Charges change depending on pH

The pI for each protein depends on the pH at which each free carboxylic acid or amino group becomes charged or uncharged, and the influence of neighbouring amino acids.

The polyacrylamide gel used in IEF is prepared commercially as a continuous pH gradient. During electrophoresis the molecules separate out when they reach a pH point in the gel which is equal to their pI – *ie* there is no charge on the molecule so it is not under the influence of an electric field. The molecules are focused in a stationary band (Figure 13).

RS•C

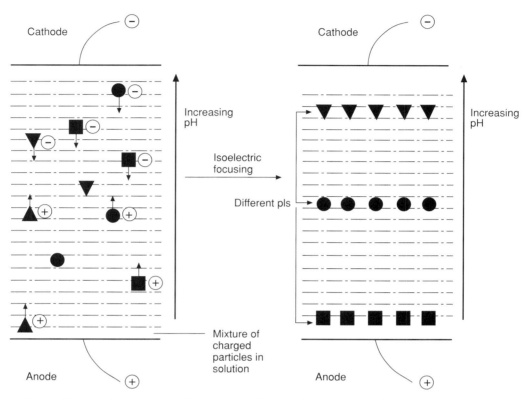

Figure 13 Left – before isoelectric focusing; Right – after isoelectric focusing

This system is most suitable for zwitterionic molecules – *ie* complex molecules such as proteins which carry both positive and negative charges.

During SDS–PAGE there is a tendency for the component molecules to diffuse into the surrounding gel once they have been resolved. During IEF, as molecules diffuse along the pH gradient away from the location of their pI, they become charged and are attracted back to the separated band so a steady state is achieved. As IEF depends on this equilibrium process, fractionation is independent of the total protein load and the time needed for electrophoresis.

Proteins from living systems can be separated because the samples need little treatment before they are run, unlike SDS–PAGE. Finally, IEF gives greater resolution than SDS–PAGE, and can separate components with pI values only 0.001 pH units apart.

Setting up the pH gradient. Charge-carrying molecules (amphoteric buffers) with closely spaced pI values are mixed with the polyacrylamide gel before it is poured. When an electric current is passed through the gel these amphoteric buffers settle at positions corresponding to their pI values, hence a pH gradient is established. The most acidic buffer moves towards the anode and the most basic moves towards the opposite end, near the cathode. The pH range across the gradient depends on the particular buffers used. The buffers are usually commercially-produced oligoamino and oligocarboxylic acids - short chains of amino acids containing a carbohydrate component. The blend and number used depends on the particular pH gradient to be established. Most practitioners use commercial preparations.

RS•C

Buffers need to have the following important properties.

■ Good acid-base buffering capacity

■ Zero mobility at their pI

■ Good conductance properties

■ Good solubility in both water and the gel

■ No reactivity with the sample

■ A lack of response to detection methods used to identify components of the mixture.

The buffers and proteins have no charge near their pIs, and they often precipitate out of solution leaving conductivity gaps. Immobilised pH gradients (IPGs) overcome this problem, by immobilising the pH gradient within the fibres of a polyacrylamide gel.

Running a sample. Isoelectric focusing can be used to identify, for example, proteins from different fish species. These can be characterised by specific proteins. Approximately equal masses of small pieces of different fish (trout, salmon, cod or coley) are homogenised. After centrifuging for 5 mins at 2000 revolutions per minute the supernatant (upper layer of liquid) is taken for spotting onto the gel.

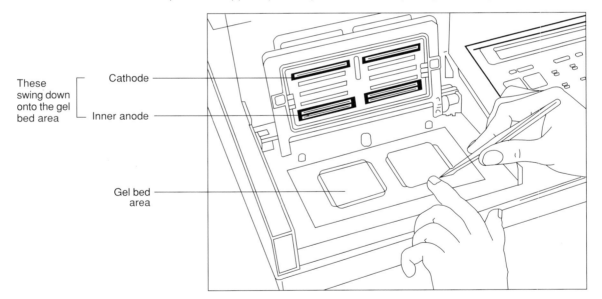

These swing down onto the gel bed area — Cathode — Inner anode — Gel bed area

Figure 14 Diagram showing gel placed in electrophoresis chamber

The applicator (a small plastic 'shark's tooth' comb) is loaded with approximately 2 μdm^3 of each sample.

A prefocusing electrophoresis is run with the amphoteric (acting as acid and base) buffers added to the gel to set up the pH gradient. The samples are then transferred to the gel. (The position of application of the samples in IEF is not crucial, as it is in SDS–PAGE, because the components migrate in an electric field to their respective pIs). Focusing then takes place by running the samples at 3.5 W and 2.5 mA for approximately 25 mins. The separated components are then stained with Coomassie Blue and destained with an aqueous solution containing 30% v/v methanol and 10% v/v ethanoic acid. An electropherogram is shown in Figure 15.

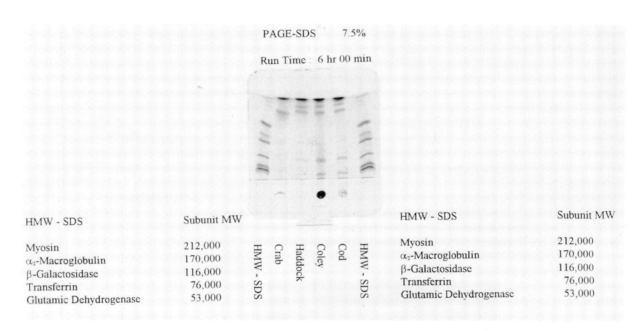

HMW - SDS	Subunit MW						HMW - SDS	Subunit MW
Myosin	212,000	HMW - SDS	Crab	Haddock	Coley	Cod	HMW - SDS	Myosin	212,000
α₂-Macroglobulin	170,000						α₂-Macroglobulin	170,000
β-Galactosidase	116,000						β-Galactosidase	116,000
Transferrin	76,000						Transferrin	76,000
Glutamic Dehydrogenase	53,000						Glutamic Dehydrogenase	53,000

Figure 15 Electropherogram of fish products

Free-solution IEF

Isoelectric focusing is an excellent method for purifying proteins. However, eluting the proteins from the gel after purification is time-consuming and can result in mixing and loss of a high proportion of the product. By carrying out IEF in free solution (*ie* without a gel), a protein mixture can be fractionated or individual components purified.

The cylindrical focusing chamber is usually 3 cm in internal diameter and 15 cm long, with maximum capacity of approximately 60 cm^3. A ceramic cooling finger along the axis of the chamber revolves at approximately one revolution per minute. The sample and buffers are added by syringe. Separations are usually run for 3–4 h at a constant 12 W. At the end of the separation the components are drawn up into separate test-tubes placed directly underneath the chamber.

The protein fractions obtained from free solution systems are purer that those obtained from IEF gel systems. Free-solution systems are suitable for preparing proteins to the purity standards required for structure determination by X-ray crystallography.

Two-dimensional gel electrophoresis

In the electrophoretic separations already described there can be problems in resolving peptides and proteins that have similar properties. For example, SDS–PAGE may not distinguish sufficiently between proteins of similar molecular mass, and IEF may not be sensitive enough to separate proteins with very similar isoelectric points. Further separations may be needed if the required proteins or polypeptides are not resolved. This can be done using two-dimensional electrophoresis, in which two consecutive separations are done on gels at right angles to each other (Figure 16).

RS•C

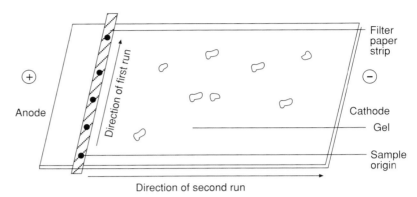

Figure 16 The basic arrangement for two-dimensional gel electrophoresis

Two-dimensional electrophoresis and isoelectric focussing can also be used in combination, for the resolution of the components of mixtures of proteins.

Applications of two-dimensional electrophoresis

With silver staining – using silver nitrate – two-dimensional electrophoresis is extremely sensitive, and is a useful technique for separating and detecting small quantities – eg 10^{-9} g – of biological samples. Two-dimensional electrophoresis is a powerful tool in research when analysing cell or nuclear proteins linked to an abnormal condition – whether genetic, metabolic or cancerous.

Defects in new born babies. Neural tube malformations can be detected from proteins leaking from the central nervous system of a foetus into the amniotic fluid (the fluid enveloping the foetus). Analysing these proteins can indicate spinal problems in new-born babies.

Sweating polypeptides. Over 400 polypeptide spots show up in a two-dimensional electropherogram of human sweat. Many of these have been previously unidentified.

Alcohol abuse. The extent of alcohol abuse can be investigated by analysing blood. This is because excess alcohol is associated with changes in acidic proteins and glycoproteins in blood plasma.

'Fish eye' disease. This is an inherited condition in which lipid is laid down in the eyes making them appear opaque like fish's eyes. Its medical name is dyslipoproteinanaemia. Protein samples from sufferers have been analyzed, suggesting that there is a deficiency in an enzyme system associated with lipid metabolism.

Heart attacks. Blood samples are taken at regular intervals (once a day for three days after initial chest pain). Studies are being done to identify more sensitive 'marker' proteins which will indicate early blockage in blood vessels.

Assessing fitness. When a person is very much out of condition there can be a serious increase in the amount of protein in urine, which can be detected by two-dimensional electrophoresis. As well as being a pointer to lack of physical fitness, increased protein content in the urine can also be a possible indicator of the onset of diabetes.

RS•C

Quantifying two-dimensional gel patterns

Over 1000 proteins can be collected and identified on a single gel in two-dimensional gel electrophoresis. Naked eye analysis of this quantity of data would be time-consuming, for this reason analysis is automated. For the results to be as unequivocal as possible any external background noise, such as background staining, is removed.

A two-dimensional analyser needs a scanning device, a means of converting and storing the data electronically, and a function to allow patterns from gels to be compared and matched up for identification and accurate analysis.

The spots appear to cover the gel in a random pattern, hence linear scanners cannot be used. Any scanner must be able to pick up information from all coordinates of the gel. The gels are scanned using a photometric measuring device, or the gel is moved mechanically over the detectors. The data are then mapped as coordinates on the gel and recorded as absorbance values, which have previously been calibrated.

Peptide mapping   **Box 6**

Whatever method of electrophoresis is used – even two-dimensional electrophoresis – there are occasions when ambiguous results are obtained. Unrelated proteins may run very similarly, while related proteins can be widely separated. For example, there might be an incomplete separation between proteins and their precursors. One example of this is the protein, thrombin, which activates the synthesis of a blood-clotting agent.

The precursor for thrombin is prothrombin, and the conversion is activated by calcium ions. If a substantial amount of prothombrin is still present this gives unreliable results on the electropherogram, because the sample does not run true for thrombin. Proteins may also be modified by phosphorylation, or degraded during sample preparation. Peptide mapping can help by giving a more detailed fingerprint – a unique identification pattern – of the make-up of specific proteins.

In peptide mapping a purified protein is broken down into its constituent peptide fragments in a specific and controlled manner. This is done using enzymic or chemical hydrolysis. The peptide mixture is then separated using electrophoresis, and the patterns of the separated zones are compared with those of standard proteins that have been treated in the same way.

Enzymic hydrolysis is more specific than chemical hydrolysis and operates in a buffer system similar to that of the proteins being investigated. However, enzymes are also proteins and will contribute peptides to the mixture. When using enzymic hydrolysis it is important to run a blank containing only the enzyme so that the peptides from the enzyme can be eliminated.

The standard technique for peptide mapping involves the following stages.

- Electrophoretic separation of the protein sample, usually through PAGE.
- Detecting different bands – *eg* by staining.
- Preparing the purified specimens by denaturing the enzymes. The sample is treated with Tris-HCl buffer, 0.5% v/v SDS, 10% v/v glycerol and 0.001% v/v bromophenol blue. The mixture is then heated at 100 °C for 2 minutes to deactivate any extraneous enzymes. The denaturation also helps in unfolding the chain to help optimum SDS bonding.
- Enzymic or chemical hydrolysis of the separated proteins.
- Stopping the hydrolysis by using 2-mercaptoethanol, SDS and boiling.
- Peptide mapping by loading in a gel slab. A better electropherogram is obtained with a concentration gradient gel.

As well as characterising protein substrates, the technique can also be used to characterise proteases – *ie* enzymes which hydrolyse proteins. An unknown protease acting on a known protein substrate cleaves bonds at specific points and produces characteristic banding as a result of electrophoresis.

RS•C

Separation techniques for deoxyribonucleic acid (DNA)

Pulsed field gel electrophoresis (PFGE)

Conventional DNA gel electrophoresis separates DNA fragments of different sizes in an agarose gel. The negatively charged phosphate backbone of DNA means that the molecule migrates towards the anode. Pulsed field gel electrophoresis is used to separate DNA fragments up to 6 Mb long. (DNA fragments are sized according to the number of bases – 6Mb is 6 million bases). In conventional DNA gel electrophoresis in agarose gel, where DNA fragments are separated according to size through a sieving effect, the size limit is 50 kb. Above this size, fragments tend to run as a broad unresolved band whatever their size. One explanation for this is that the DNA molecules become stretched out under high field strengths, and thus the effect of increased molecular mass becomes diminished.

Pulsed field gel electrophoresis (Box 7) increases by two orders of magnitude the limits on the size of DNA molecules that can be analysed and separated. This technique can be used to determine the size of chromosomes, and to separate fragments in non-bacterial cells – *eg* in mammals, where the sizes of DNA fragments that resolve into genes exceed 50 kb. In organisms such as bacteria and yeasts where the individual genomes and chromosomes cannot be resolved through microscopy, PGFE provides a novel method of identifying these chromosomes.

How it works

The DNA is loaded near the cathode, and travels towards the anode through a concentrated agarose gel, migrating under the influence of two electric fields which are held at an angle greater than ninety degrees. These fields are regularly pulsed – *ie* alternated in relation to each other without any 'dead-time' in the alternation. At high voltage gradients the DNA molecules are stretched along the direction of the field in an effort to penetrate the pores of the gel and to make a net forward movement .

When the field is pulsed, the DNA molecules have to reorientate and migrate at a different angle to the direction at which they are stretched. The longer the DNA molecule, the longer it takes to reorientate, and longer molecules are held back in the gel relative to the shorter ones. The resolution depends on the time interval between pulses.

RS•C

Mechanism   **Box 7**

A theoretical explanation of PFGE has not been established although intuitive models have been elaborated from models for conventional DNA electrophoresis in agarose gel. The 'reptation' model suggests that at high field strengths the DNA molecule stretches out along the line of direction of the field and coils around obstacles and through pores much like a slowly moving snake (Figure 17). This model is consistent with the observation that the rate of migration of DNA is independent of molecular mass.

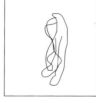
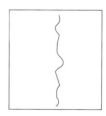

Low field   Moderate field   High field

Figure 17 A schematic diagram showing the elongation of DNA under the influence of an electric field

The migration of DNA molecules in PFGE has been monitored by staining DNA from the yeast *S.cerevisiae* with fluorescent dye and following the movement of molecules through a microscope mounted over the electrophoresis chamber. The molecules extend as they move lengthwise along the field. When the field direction changes both ends of the molecule may start off in the new direction with the result that the molecule adopts a U-shaped configuration. However, in most cases it appears that the previously trailing end takes over the leading position as the molecules slide off in the new direction. The fluorescence increases at the leading edge, suggesting that the molecule bunches when encountering obstacles.

Very high field strengths appear to impede the DNA's movement and resolution. This may be because the DNA is uniformly charged along its length because of its phosphate backbone. A large molecule might be forced to move simultaneously through separate pores leading to trapping and entanglement. This effect increases as the field is constantly switched.

Using restriction enzymes

Restriction enzymes – endonucleases – are used to cleave double-stranded DNA molecules at specified 4–6 base pair sites – producing fragments. These can then be used for cloning eg human DNA can be cloned into the E.Coli gene to synthesise a protein that helps to destroy cancerous tumour cells. Restriction enzymes can also be used for genetic mapping. By cutting the human chromosome into manageable segments a base map of the DNA can be constructed and sequences for gene expression can be located.

The properties of restriction enzymes – which are known to cleave the DNA of a particular species at particular points – can be used for calibration in PFGE. Given lengths of DNA a known number of base pairs can be run through PFGE and identified.

One of the problems in using restriction enzymes is the difficulty in cleaving segments of an appropriate length in mammalian DNA. Some of the segments are too long for mapping while others are too short – *ie ca* 4–8 base pairs in length, which makes it difficult to analyse these segments for gene expression.

Pulse time and field strength

Increasing the pulse time from say 4-second pulses to 60-seconds pulses leads to progressively greater resolution of larger fragments of DNA to a maximum of *ca* 1600 kb.

RS•C

With all other conditions remaining constant, longer pulse times give the resolution of larger fragments. However, there is a limit to the length of pulse time and the size of fragment. Trapping can occur with very long segments and a lower field strength must be used. Even longer pulse times are needed leading to very long slow gel runs – typically 24–48 hours at 1–2 V cm^{-1} with a 30 min pulse time.

The polymerase chain reaction (PCR)   **Box 8**

The polymerase chain reaction was discovered and patented by Kary Mullis in 1983 (Mullis shared the Nobel prize for chemistry in 1993 for his discovery). It is a method for copying DNA fragments (Figure 18), so that up to a hundred billion copies of a section of DNA can be made in a few hours from a single molecule. Because of this simple, quick method and its extremely large amplification, it is now possible to analyse minute quantities of DNA. This technique can be used to study samples of DNA from fossils – *eg* dinosaurs, hairs at crime scenes, and to investigate the geographical spread of disease carriers, such as the screwworms, by comparing their DNA patterns in different parts of the world.

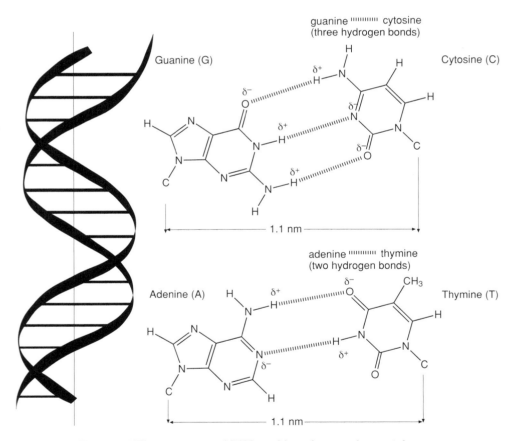

Figure 18 The structure of DNA and how base pairs match up

An individual segment of a DNA molecule is extracted. The temperature is raised to 90 °C which separates the individual strands of the DNA duplex. The temperature is lowered to ca 40 °C and synthetic DNA fragments – oligonucleotides consisting of small numbers of nucleotides arranged in a specific order – are added. These bind to the strands at the correct positions. The temperature is raised to about 70 °C and the enzyme DNA polymerase, which is added, builds up two new complete copies of the DNA strands. By repeating this cycle of heating and cooling, the strands are separated and built up again. The process works like a photocopying machine with millions of copies produced in a short time (Figure 19).

RS•C

Box 8 continued

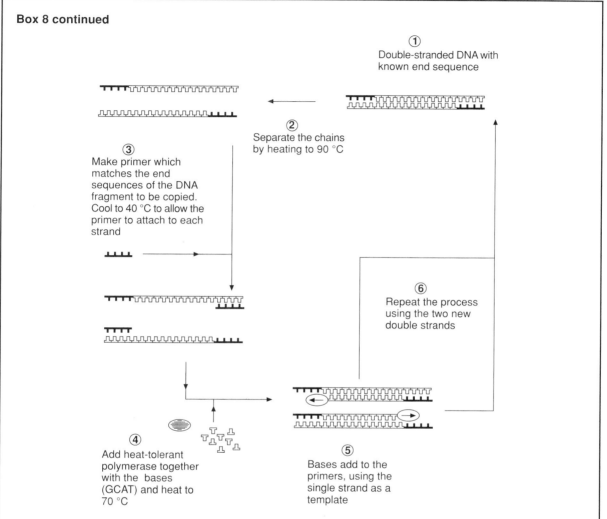

Figure 19 The polymerase chain reaction (PCR)

Although most enzymes are deactivated by excessive heat, DNA polymerase was first extracted from the thermophilic bacterium *Thermus aquaticus* (Taq) and can withstand heating. The enzyme is produced by genetically modified organisms. The DNA fragments can then be electrophoresed on agarose, blotted (see below) and studied further.

Blotting

Blotting involves transferring separated components from the electrophoresis gel to a thin support membrane, most commonly nitrocellulose or nylon. The membrane – much easier to handle than the gel – binds and immobilises the components. The details of the different blotting techniques vary according to the manufacturer but the two most commonly used are western blotting or electroblotting, mainly for proteins and polypeptides and southern or capillary blotting for DNA and RNA. Membranes are graduated according to pore size, those with smaller pore sizes being able to retain smaller fragments.

RS•C

The advantages of the thin nylon support are:

■ the processing times for detection are much shorter; and

■ successive analyses – *eg* multiple assays – can be performed, and the blots can be stored for long periods of time.

In both blotting methods the membrane and the gel are in close contact under pressure. The electrodes are encased in plastic holders and are designed so that a uniform electric field is applied across the gel and the matrix (Figure 20). A buffer is used to act as a suitable medium for the transfer.

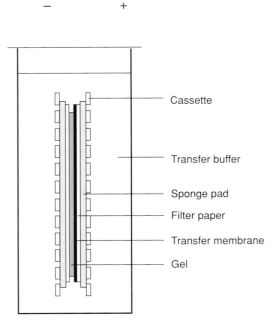

Figure 20 A schematic diagram for electroblotting

Capillary blotting or Southern blotting (Figure 21) similarly involves compressing the gel and the membrane, forcing the components through the pores of the interconnecting filter paper and on to the membrane. However, an electric field is not used.

RS•C

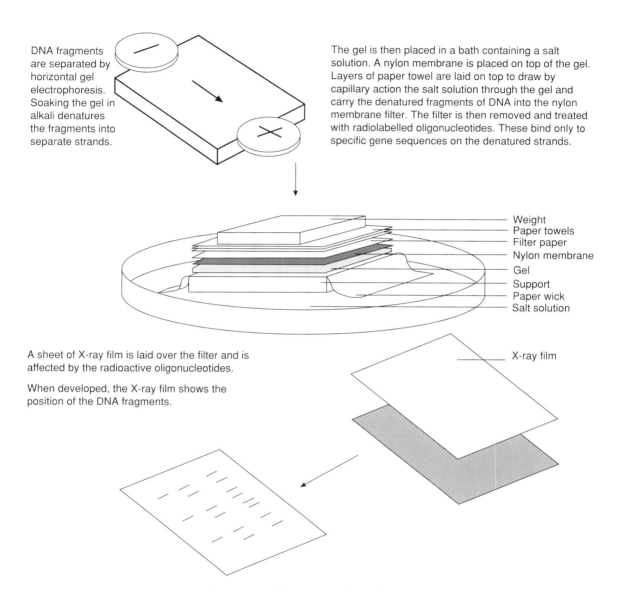

DNA fragments are separated by horizontal gel electrophoresis. Soaking the gel in alkali denatures the fragments into separate strands.

The gel is then placed in a bath containing a salt solution. A nylon membrane is placed on top of the gel. Layers of paper towel are laid on top to draw by capillary action the salt solution through the gel and carry the denatured fragments of DNA into the nylon membrane filter. The filter is then removed and treated with radiolabelled oligonucleotides. These bind only to specific gene sequences on the denatured strands.

Weight
Paper towels
Filter paper
Nylon membrane
Gel
Support
Paper wick
Salt solution

A sheet of X-ray film is laid over the filter and is affected by the radioactive oligonucleotides.

When developed, the X-ray film shows the position of the DNA fragments.

X-ray film

Figure 21 Capillary (or Southern) blotting

Detection

The electrophoretic run enables the separation of the components of a mixture. However, these then have to be characterised or identified. For most identification techniques the gel has to be removed from the gel holder. The ease with which this can be done depends on the gel (agarose is mechanically stronger than polyacrylamide) and, for polyacrylamide, the concentration of acrylamide monomers – polyacrylamide gels with low concentrations of acrylamide are almost liquid. For weak gels identification techniques which can be done without removing the gel should be used. Alternatively, a blot can be done and the analysis carried out on a membrane such as nitrocellulose or nylon.

Staining

The separated components (bands) are reacted with a stain or a dye. Their positions are indicated by production of a colour. Some staining techniques produce a depth of colour depending on the amount of component present, which can then be quantified

RS•C

by absorption or fluorescence. The stains must have very specific properties and are chosen relative to the chemical nature of the components. They must be:

■ selective and act only on the components that need to be detected;

■ stable so they do not fade quickly;

■ able to react rapidly with the component; and

■ insoluble in the liquids used for destaining.

Destaining (Box 9) is necessary to remove any unreacted stain, thereby defining the component's location more precisely. For quantification, the stained compound should be susceptible to absorption measurements or to fluorescence emission. Staining can be done before or after the electrophoresis run, depending on the nature of the sample, the stain used and the purpose of the separation – *eg* where the electrophoretic separation needs to be followed visually.

Protocols for staining vary, depending on the range of factors listed above, in addition to gel thickness and concentration. However, most stains work adequately if they are left overnight and destaining is done the following morning.

Destaining   **Box 9**

There are two main methods for removing unbound stain - diffusion destaining and electrophoretic destaining.

Diffusion destaining. In this method the gel is immersed in a destaining solution and the unbound stain leaches out into the solution. For stains such as Coomassie Blue (used for staining proteins) the destaining solution is normally a solution of 12.5% v/v 2-methylpropan-1-ol and 10% v/v ethanoic acid. The destaining solution is periodically renewed before it becomes saturated. Commercial destainers often contain charcoal for adsorbing the dye.

Electrophoretic destaining. After staining, an electrophoresis is done at right angles to the original separation. The unbound molecules of the dye migrate into a solution of 7% v/v ethanoic acid (Figure 22).

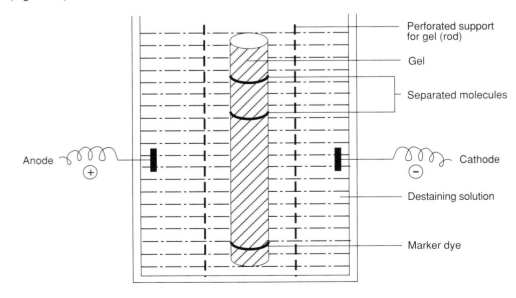

Figure 22 Apparatus for electrophoretic destaining

RS•C

Dyes

There are many commercial dyes used in detection. One of the most popular commercial stains for protein detection is the Coomassie Blue range. This is a good stain for qualitative work, but equal amounts of different proteins bind the stain unequally, so quantitative staining requires a standard curve for a particular protein.

Silver staining – using silver nitrate solution – is up to a 100 times more sensitive than Coomassie Blue and can detect as little as 10^{-18} g after electrophoresis.

However, silver staining has a number of disadvantages, and should only be used where its increased sensitivity is definitely required. The main problems are that silver staining:

- is expensive;

- is laborious to do;

- leaves a high background stain;

- is fairly unselective (it stains polysaccharides and DNA, and proteins); and

- in some cases stains poorly.

A different range of stains is used for nucleic acids. One of the most popular is ethidium bromide (2,7-diamino-10-ethyl-9-phenyl-phenanthridinium bromide) which is fluorescent and can be seen using ultraviolet radiation.

Quantification methods

There are two main methods for estimating the amounts of different components after staining and destaining. These are densitometry, based on light absorption using adapted conventional spectrophotometers, and fluorescent emission.

Densitometry. After destaining, a track is cut from the gel and is scanned through a narrow fixed parallel light beam. This process is called linear tracking. The intensity of light transmitted is detected by a photomultiplier and computerised to provide a read-out of the different amounts of components.

Fluorescent emission. The advantage of fluorescent emission is that the progress of the electrophoresis can be followed. The detection instrumentation is similar to that used for densitometry. Before electrophoresis, the protein is stained with a fluorescent dye such as fluorescamine for proteins and ethidium bromide for nucleic acids, and then subjected to ultraviolet light in a darkened room.

Radiochemical detection methods

Radiochemical detection methods are much more sensitive than staining. The sample is labelled with a radioactive isotope before electrophoresis, and the detection method essentially depends on the labelling technique used. There are two main techniques: the more sensitive autoradiography where an image is scanned onto an X-ray film, and liquid scintillation counting (LSC) which involves dissolving the component bands from the gel. This means slicing the gel – hence LSC is not used for soft gels.

Autoradiography. Radioactive emissions are captured on an X-ray film that can then be scanned with a densitometer. Often the gel is stained and destained so that the photographic image can be compared with a stained banding pattern. After labelling and electrophoresis the gel is placed against an X-ray film. This can be done using a spring-loaded cassette to ensure close contact. The time of exposure depends on the label used, the amount of radioactivity in each band, and whether the analysis is intended to detect and quantify major components or as many components as possible.

RS•C

Liquid scintillation counting. This is particularly good for weak emitters (Box 10). It is not as sensitive as autoradiography and involves the elution of components from the gel before measurement.

Before counting, the gels have to be sliced and dissolved. Agarose gels can be sliced using sharp razor blades, but polyacrylamide gels are more easily distorted so specialised slicers – eg Mickle gel slicers – need to be used.

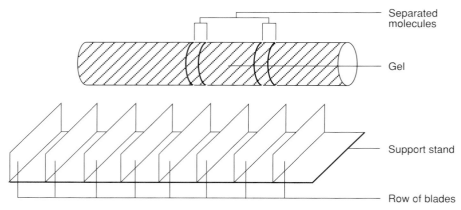

Separated molecules

Gel

Support stand

Row of blades

Figure 23 Apparatus for slicing polyacrylamide tube gels

The components can either be dissolved in commercial solubilisers or the gel structure can be broken down and the components leached into solution. Methanamide, $HCONH_2$, is used to break down agarose gels and 30% v/v hydrogen peroxide is used for polyacrylamide gels. The liquid is then put into a sample chamber for counting.

Weak emitters   **Box 10**

Low energy α-particles from weak emitters such as tritium (^3H) are unlikely to penetrate the gel and reach the film. One way of overcoming this problem is to impregnate the gel with a scintillator. The α-particles interact with the scintillator, generating visible light energy which forms an image on blue-sensitive X-ray film. This system needs to be run at a low temperature (between -20 °C and -70 °C). The low temperature is needed to stabilise image formation during long exposures to light.

The main chemical scintillators, or fluorophores, are 2,5-diphenyloxazole (PPO) and sodium salicylate ($HOC_6H_4CO_2Na$). There are also a number of commercial fluorography agents. The advantage of sodium salicylate is that it is cheap, water-soluble and easy to use. However, the film image produced from it is not as sharp as that from using PPO.

There are two problems that need to be taken into account when using fluorophores. First, staining diminishes the intensity of the film image probably by quenching the fluorescence. Secondly, there is a difficulty with concentration gradient polyacrylamide gels because of a tendency to absorb the fluorographic image.

Biological assay

Enzymes. Proteins such as enzymes can be detected by their biological activity. For example, enzymes that hydrolyse esters can be detected by the following reaction:

$$\text{1-Napthylethanoate} \xrightarrow{\text{esterase}} \text{1-Napthol + Ethanoic acid} \xrightarrow{\text{Fast red}} \text{Coloured compound}$$

RS•C

When detecting enzyme activity it is important that the sample is not denatured during detection and quantification. The electrophoresis is done:

■ at a pH which does not denature the protein; and

■ in the cold – *ie* there should be an efficient cooling system to dissipate excess heat.

The particular advantage of a biological assay is that it is extremely selective and sensitive – selective because enzymes act on specific substrates, and sensitive because they react in small amounts. The enzymes can be assayed either by gel slicing followed by elution into a buffer, or by staining.

The detection technique chosen depends on the particular class of enzyme. For example, dehydrogenases can be localised by incubating the gel in a solution of tetrazolium salt to give a coloured product.

Immunoelectrophoresis. A variety of techniques have been devised in which electrophoresis is combined with the ability of proteins to act as antigens (Figure 24), producing antibodies – *ie* an immune response.

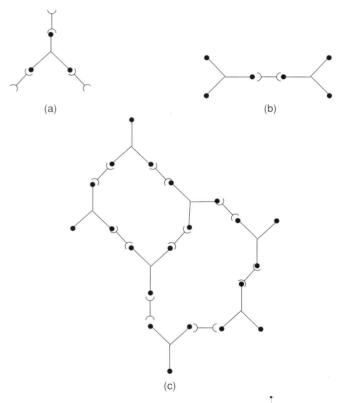

(a)

(b)

(c)

Figure 24 Reactions of antibody (>—<) with trivalent antigen (•⁄\•) in conditions of (a) antibody excess, (b) antigen excess and (c) at equivalence point

A protein can be detected by reacting it with an equivalent amount of antibody resulting in the formation of precipitate, called precipitin (Figure 25). The time for antigen-antibody link-ups needs to be monitored carefully for individual systems, and particularly for the species in which the antigens for the antibodies is raised. For example, precipitin from horse antibodies tends to redissolve in excess antigen or antibody while rabbit antibodies produce much more stable antigen-antibody crosslinkages. Immunoelectrophoresis can be used for proteins that do not show enzymic activity. The procedure is carried out in agarose gels.

RS•C

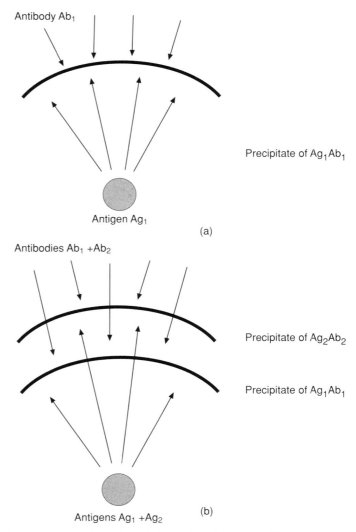

Figure 25 Formation of precipitates in gel immunodiffusion (a) monospecific anti-serum; (b) bispecific antiserum. The rate of diffusion of the antigens is proportional to their molecular sizes, and therefore discrete precipitin lines can be produced as in (b)

Immunoelectrophoresis elicits the response of a specific antibody. It is particularly useful for analysing small amounts of individual proteins in mixtures, and for checking the purity of protein preparations by confirming the presence of only one antigen.

There are three main techniques used in immunoelectrophoresis. Classical immunoelectrophoresis is primarily qualitative and is used for identifying components in mixtures, while crossed and rocket immunoelectrophoresis are suitable for quantitative determinations.

Classical immunoelectrophoresis. In this system the proteins are separated by electrophoresis and then allowed to diffuse into the gel (Figure 26).

RS•C

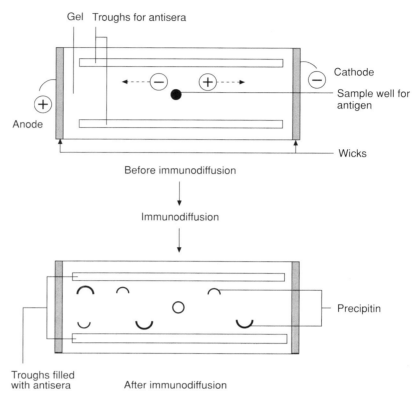

Figure 26 Classical immunoelectrophoresis

Troughs filled with antiserum are placed parallel to the electrophoresis run. The antibodies also diffuse into the gel after electrophoresis, and precipitin forms. The antigens tend to have a smaller molecular mass than the antibodies, so they diffuse faster and tend to form leading edges when they react with the antiserum.

The precipitin lines can be enhanced through staining. The ideal stain should be aqueous, leave little or no background after washing and have good photographic reproduction.

Crossed immunoelectrophoresis. Like classical immunoelectrophoresis this is a two stage technique. The first stage is the same as in the classical immunoelectrophoresis. The second stage involves an electrophoresis at right angles to the first run, towards a gel containing an antiserum. If the antibodies are in the correct concentrations a rocket shaped precipitate is formed. This technique is advantageous in that the area enclosed by the precipitate is proportional to the amount of antigen, and up to 30 proteins can be resolved and quantified in each run.

Rocket immunoelectrophoresis. This is a one-dimensional technique enabling the determination of proteins in body fluids without prior purification. The gel is impregnated with a 1% concentration of antibody. Wells are cut in the gel close to the anodic end, and precise volumes of antigen of unknown concentration and hence amount, are placed in each well. Standards – *ie* antigens of known concentration are also applied.

In an electric field, the protein samples migrate towards the cathode. At the start of the migration there is excess antigen relative to the antibody concentration, and so little precipitin forms. When the concentration of the leading edge of the antigen is equivalent to the antibody present the precipitin forms preventing further migration.

RS•C

This results in long rocket-like streaks (Figure 27). The height of the streak is proportional to the amount of the specific protein present. The standard concentrations enable a calibration curve to be drawn of concentration of protein against height, and thus the amounts of the unknown proteins can be estimated.

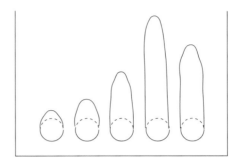

Figure 27 The rocket electrophoresis of albumin in antibody containing agarose gel (Amounts from left to right are 1,2,4,8,6 µg)

Applications

DNA fingerprinting

Gel electrophoresis is used to separate and identify DNA fragments for 'fingerprinting', for example in establishing family relationships and as evidence in criminal cases. Samples of DNA can be obtained from white blood cells, inner cheek cells or hair can be compared with those of a possible relative or those suspected of committing a crime.

Deoxyribonucleic acid (DNA) is a macromolecule with the familiar shape of a double helix, and is the main chemical species in chromosomes where the genes that determine heredity are located. The molecule consists of a sequence of base pairs (see Figure 18), the sugar deoxyribose and a phosphate backbone. It is the particular sequence of base pairs that acts as the template for the series of cellular reactions that control heredity in developing organisms.

Only about 10% of the total length of the human DNA molecule acts as a coding agent or template for those chemical reactions that result in protein synthesis. This sequence of base pairs comprises the genes. This sequence is relatively stable and is uniform within a particular species. If gene structure were unstable – *ie* changed from generation to generation – faulty protein production would result.

By far the largest proportion of DNA consists of non-coding regions where there is more variability between individuals. There are sequences of base pairs (10-60 base pairs in length) in the non-coding regions that are repeated many times. All humans have these repeats but what varies between individuals is the number of these repeats. The name given to these is 'variable number of tandem repeats' (VNTR). These VNTRs can also occur within the non-coding regions of active gene sites. On the gene that codes for myoglobin – the protein that controls oxygen concentration in muscle cells – there is a non-coding repeat sequence of 16 base pairs.

The numbers of these repeats are unique to each individual, and can act as a genetic fingerprint. Relationships, such as paternity, can be determined because this sequence is inherited. An individual receives half the non-coding sequences from the mother and the other half from the father.

RS•C

Making a DNA fingerprint

When making a DNA fingerprint the first step is to extract the DNA from a sample such as white blood cells. (All human cells contain DNA except for red blood cells). Where the DNA is in short supply it can be amplified using the polymerase chain reaction (Figure 19). Probes (Figure 28) are used to detect selected fragments of DNA because they contain sequences complementary to the VNTRs.

DNA probe

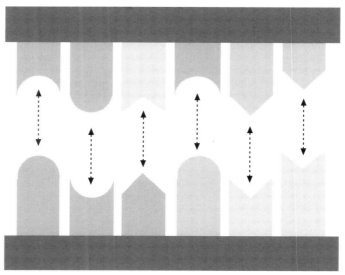

Complementary sequence

Figure 28 Diagram to show how probes work

For example, the probes are tagged with chemiluminescent molecules that give off light when exposed to X-rays.

Figure 29 shows the sequence of events in DNA fingerprinting.

RS•C

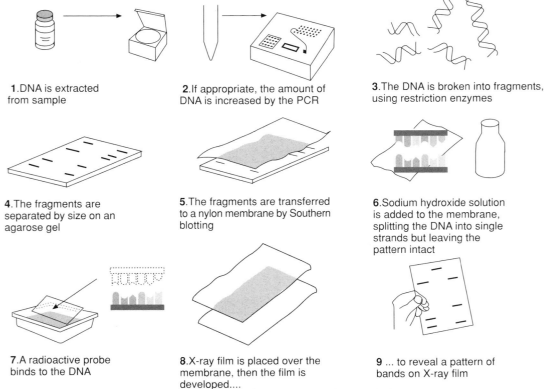

1.DNA is extracted from sample

2.If appropriate, the amount of DNA is increased by the PCR

3.The DNA is broken into fragments, using restriction enzymes

4.The fragments are separated by size on an agarose gel

5.The fragments are transferred to a nylon membrane by Southern blotting

6.Sodium hydroxide solution is added to the membrane, splitting the DNA into single strands but leaving the pattern intact

7.A radioactive probe binds to the DNA

8.X-ray film is placed over the membrane, then the film is developed....

9 ... to reveal a pattern of bands on X-ray film

Figure 29 The sequence of events in DNA fingerprinting

Fragmenting the DNA. The DNA is cut into fragments by so-called restriction enzymes (Figure 30 or Box 11). A restriction enzyme that cuts the DNA into sections at specified regions is selected. For fingerprinting, restriction enzymes that snip the DNA outside the non-coding regions are used.

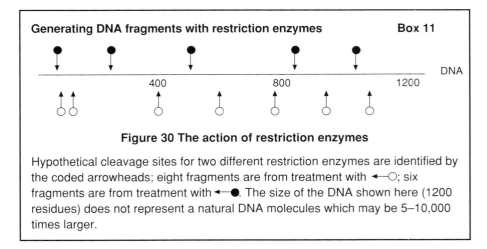

Generating DNA fragments with restriction enzymes   **Box 11**

DNA

400   800   1200

Figure 30 The action of restriction enzymes

Hypothetical cleavage sites for two different restriction enzymes are identified by the coded arrowheads; eight fragments are from treatment with ◄─○; six fragments are from treatment with ◄─●. The size of the DNA shown here (1200 residues) does not represent a natural DNA molecules which may be 5–10,000 times larger.

Electrophoresis of the fragments. The fragments are separated on an agarose gel according to their size. The sample wells are cut at the cathode and, because DNA has a negative phosphate backbone, fragments move towards the anode when the electric field is applied. The separated fragments are then treated with alkali to uncouple the duplex into single strands.

RS•C

Blotting. The single stranded fragments are blotted onto a nylon membrane and bathed in the appropriate probes – *ie* pieces of radioactively-labelled DNA containing base sequences complementary to the fragments. Excess or unwanted probes can be washed off.

Detection. The labelled DNA fragments are detected using, for example, X-ray film. This reveals a distinct pattern of fragments known as a fingerprint.

DNA offprints from newspapers

Paternity testing

After fragmentation, electrophoresis and blotting, the DNA fragments from an individual are hybridised using selective probes. Paternity testing (Figure 31) is based on the principle that a child acquires one set of tandem repeats from each of its parents – two fragments containing these repeats are detected by the probes. The DNA pattern from a child, set against those of true parents, will show markers that reflect this relationship.

M=mother
F=father
C1-C4=child

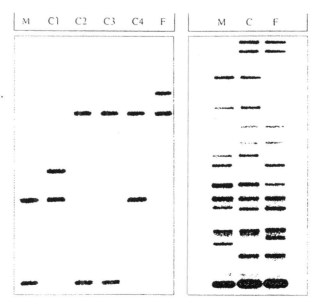

Figure 31 Paternity testing. Left shows a family with four children. F is the father of C2, C3 and C4 but not C1. Right shows every band in the child's DNA is present in either its mother's or its father's pattern.

RS•C

Problems with DNA testing

With the establishment of a DNA database, many questions have been raised about the reliability of DNA fingerprints. There is a one in 200 million chance of two individuals having the same DNA banding pattern. However, this figure is disputed, some arguing that the chances are much greater particularly for convictions in criminal cases, or for politically sensitive issues, such as identifying family relationships for the purposes of immigration control. There are some issues associated with DNA testing;

■ Evidence in a crime investigation could be mistreated.

■ Samples of DNA taken from materials for forensic evidence can be degraded by bacteria if the sample is not kept under the proper conditions.

■ Samples of DNA may become contaminated from extraneous sources if proper protocols for collecting evidence are not strictly adhered to.

■ The material substrate from which the sample is taken can sometimes affect banding patterns. The bands on a gel may be indistinct, so an inference based on these would not be possible.

■ Database evidence could be misused. Will organisations such as banks and insurance companies have access to these data? What controls will be enforced?

There is the question of whether DNA fingerprints are really unique, or are there closer relationships of VNTRs within certain population groups?

Should DNA fingerprints be used as evidence of family relationships for immigration control?

There is a concern about the Human Genome Project in that people may be scanned for 'at risk' genes and then be eliminated from services such as mortgages, life assurance and jobs.

RS•C

RS•C

Problems

1. The table below shows a number of proteins and their isoelectric points (pI).
   Complete the table, deciding what charge each protein will carry at the pHs listed.

Protein	pI	pH 4.0	pH 7.7	pH 10.0
Triose phosphate isomerase from rabbit muscle	6.8			
Adenine phosphoribosyl transferase from human blood cells	4.8			
Lysozyme	11.1			
Human haemoglobin	7.1			
Insulin	5.4			
Cytochrome c	10.0			
β-lactaglobulin from cow serum	5.2			
Ceramide trihexosidase from human plasma	3.0			
Pepsin	1.0			

2. Suggest the pH of the buffer medium which would be best at separating
   cytochrome c from haemoglobin.

Answers

1.

Protein	pI	pH 4.0	pH 7.7	pH 10.0
Triose phosphate isomerase from rabbit muscle	6.8	+	-	-
Adenine phosphoribosyl transferase from human blood cells	4.8	+	-	-
Lysozyme	11.1	+	+	+
Human haemoglobin	7.1	+	-	-
Insulin	5.4	+	-	-
Cytochrome c	10.0	+	+	zero
β-lactaglobulin from cow serum	5.2	+	-	-
Ceramide trihexosidase from human plasma	3.0	-	-	-
Pepsin	1.0	-	-	-

2. pH of 7.7

RS•C

This page has been intentionally left blank.

Capillary electrophoresis

Introduction

Capillary electrophoresis (CE) (Figure 1) includes a range of techniques in which an electric potential is applied along a length of capillary tube made of silica (internal diameter 10×10^{-6} m to 100×10^{-6} m, and up to 1 m long) to separate different components within a buffered mixture. Types of analyte range from large biopolymers, such as proteins and deoxyribonucleic acid (DNA), down to small anions and cations.

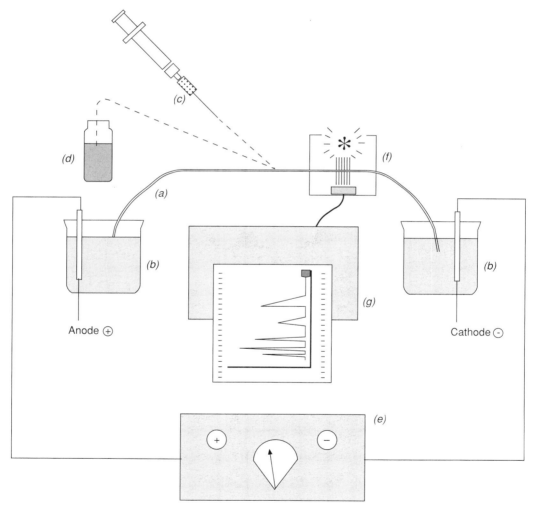

Figure 1 The basic components of a capillary electrophoretic system. *(a)* Fused silica capillary; *(b)* electrolyte vessels with electrodes; *(c)* syringe-to-capillary adaptor (replaced in commercial instruments by pressure or vacuum-driven rinse); *(d)* sample vial raised to a level necessary for sample introduction by hydrostatic pressure; *(e)* regulated high voltage power supply; *(f)* detector; *(g)* data acquisition device (recorder, integrator, or computer).

RS•C

The capillary tube is filled with a solution containing different chemical components. When a voltage is applied this acts as a driving force to separate the components into discrete bands or zones (Figure 2).

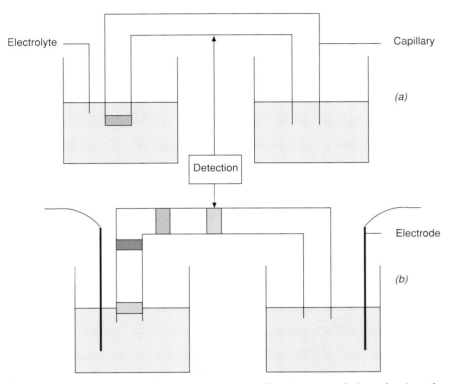

Figure 2 Two initial stages of CE separation *(a)* **Following a sample introduction, the capillary is transferred back into the electrolyte container for application of a separation potential;** *(b)* **the separation potential has been applied for a time period coinciding with the migration time of the first of the several analyte zones**

The mobility of a species due to its charge in the electrical field is countered by its 'drag' through the medium of the buffer, hence the species separate on the basis of their mass-to-charge ratio. (This is also discussed in the chapter on Gel electrophoresis.) As each zone passes a detector, a response is recorded.

The large surface-to-volume ratio of the capillary tube enables heat to be dissipated so that the system can operate at high voltages, giving fast and efficient separations, requiring only small amounts of material – eg 10^{-9} g.

Capillary electrophoresis has several additional attributes including:

■ the ability to separate substances varying largely in ionic/molecular size;

■ complete automation;

■ no problems with disposal of solvent (amounts are very small);

■ low running costs; and

■ simple experimental procedure.

RS•C

Principles and instrumentation

In CE analysis, an appropriate buffer is used which can best separate the components. Selecting the buffer depends on knowledge of the chemistry of the sample components and previous experience of using CE. This is often achieved by trial and error. The criteria for using certain buffers are discussed in Box 1. The solution to the problem at the end of the chapter also discusses the effects of using different buffers.

Analysis time is shorter with high field strengths, hence the heat generated by high voltages must be rapidly dissipated (Figure 3). All commercial capillary tubes are made of silica which is a good heat conductor. A polyimide coating on the outside protects the brittle capillary tube.

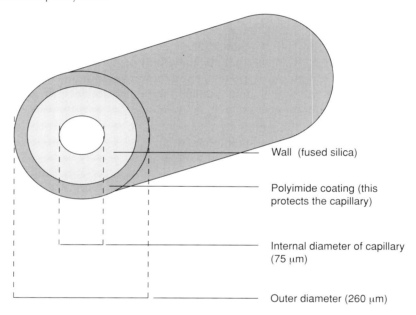

Wall (fused silica)

Polyimide coating (this protects the capillary)

Internal diameter of capillary (75 μm)

Outer diameter (260 μm)

Figure 3 A cross-section of a polyimide-coated fused silica capillary

Other ways of avoiding overheating are to use low conductivity buffers and efficient capillary cooling. Most capillary tubes are incorporated into cartridges which are air-cooled.

There are two types of flow in CE.

- **Electrophoretic flow.** In this type of flow positively charged species migrate towards the cathode and anions migrate to the anode in response to the potential difference generated across the length of the tube.

- **Electroosmotic flow (EOF) (Box 1).** This flow results from the interaction between the ionic charges on the internal wall of the capillary tubing and the polar charges residing in the bulk liquid. There are two important effects.

1. The leading profile of the moving column of the liquid is flat, resulting in band-sharpening because the same components move together in narrow zones.

2. When the EOF has a greater force than the electrophoretic flow all ions, irrespective of their charge, are swept in the same direction. This is an advantage for detection because each zone passes a detector window in turn (Figure 4), hence a property such as absorbance can be recorded against migration time.

RS•C

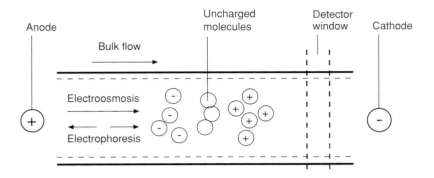

Figure 4 Schematic representation of electrophoresis and electroosmosis in a separation of anionic, neutral and cationic analytes. All ions and molecules are swept in the same direction by electroosmosis in the order: highly charged cations > cations > neutral molecules > anions > highly charged anions

Box 1 follows overleaf.

RS•C

RS•C

Explaining electroosmotic flow (EOF)   **Box 1**

Electroosmotic flow in CE results from the interaction between the ionic charges on the internal wall of the capillary tubing and the polar charges in the bulk liquid. The inner surface of the capillary tube is made up of silanol groups (Si–OH), from the silica, which can be ionised when in contact with the surrounding buffer. The isoelectric point of silica is ca pH 1.5, (which is rarely used) – *ie* it is uncharged at this pH. This means that the silanol groups are negatively charged at pH values above 1.5. A buffer such as sodium borate solution at a concentration of 2×10^{-2} mol dm^{-3} has a pH of 9, hence the negatively-charged silanol groups attract positive ions from the buffer forming an electrical 'double layer' called the Helmholtz layer (Figure 5).

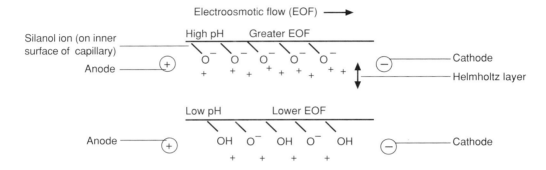

Figure 5 A schematic representation showing the arrangement of the silanol and Helmholtz layers. A high pH produces a greater degree of ionisation in the silanol layour, leading to a greater EOF. Buffers with higher pHs increase EOF.

The Helmholtz inner layer consists of fixed negative charges and the outer layer has water molecules precisely oriented towards the silanol groups, and a looser array of positively-charged hydrated ions. When a potential is applied across the length of the capillary tube the positively-charged hydrated ions in the outer diffuse region of the Helmholtz layer migrate towards the cathode taking the associated water molecules with them and generating a bulk flow of water within the capillary. As the hydrated ions leave the vicinity of the wall, more take their place, maintaining the EOF.

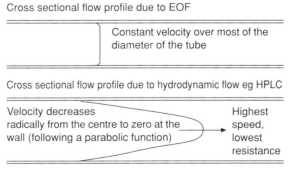

Figure 6 Comparing a flat profile with a parabolic profile – which arises from pressure-driven systems. Parabolic profiles give band-broadening and less defined peaks

With the buffering conditions of borate (2×10^{-2} mol dm^{-3}) at pH 9, the EOF speed is about 2×10^{-3} m s^{-1}. For a capillary tube with an internal diameter of 5×10^{-5} m this translates to a volume flow of 4×10^{-9} dm^3 s^{-1}. Buffers with a higher pH increase the EOF speed because they increase the ionisation of the silanol groups. The rate of flow is measured by injecting a neutral solute such as phenylmethanol ($C_6H_5CH_2OH$) at the sampling end near the cathode and measuring the time it takes to reach the detector. Increasing the flow speed may be desirable for speed of separation but high pH buffers may have an adverse affect on some separations – *eg* by reducing the positive charge on some species. Choosing the correct buffer means compromising between rate and direction of flow, and degree of selectivity. The leading profile (Figure 6) of the moving column of liquid has a flat leading profile – plug-flow – leading to high efficiencies (Box 2). This is due to the EOF originating at the wall of the capillary tube. This contrasts with the parabolic profile generated by pressure-driven systems such as High Pressure Liquid Chromatography (HPLC).

RS•C

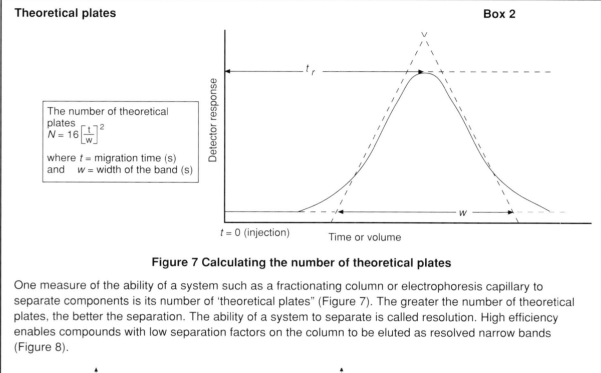

Theoretical plates   **Box 2**

The number of theoretical plates
$$N = 16 \left[\frac{t}{w} \right]^2$$

where t = migration time (s)
and w = width of the band (s)

$t = 0$ (injection)

Figure 7 Calculating the number of theoretical plates

One measure of the ability of a system such as a fractionating column or electrophoresis capillary to separate components is its number of 'theoretical plates" (Figure 7). The greater the number of theoretical plates, the better the separation. The ability of a system to separate is called resolution. High efficiency enables compounds with low separation factors on the column to be eluted as resolved narrow bands (Figure 8).

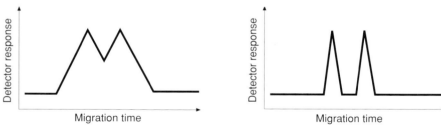

Figure 8 The influence of peak efficiency in a difficult separation

Preventing electro-osmotic flow

In some CE techniques, such as capillary isoelectric focusing, the separated components have to remain stationary, and EOF must be prevented. This can be achieved by altering buffer conditions, using additives – eg cetyltrimethylammonium bromide ($CH_3(CH_2)_{15}N(CH_3)_3Br$, CTAB), or coating the capillary tube wall with a polymer such as methyl cellulose.

One problem with separating polar chemicals is adsorption on the walls of the capillary tube through electrostatic interactions. The problem is partly overcome by using coatings with hydrophobic properties to reduce these interactions.

Some cations co-elute – ie migrate at the same speed in uncoated capillary tubes. Coating the capillary walls enables two cations to be specifically resolved which may be undifferentiated under normal conditions.

Introducing the sample

The two most common techniques of introducing a sample into a capillary tube are through a pressure differential or by electromigration.

RS•C

Pressure differential. This is the most common method of introducing a sample in CE, and there are three variants (Figure 9).

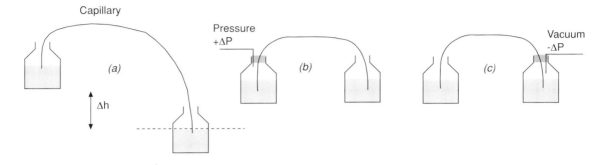

Figure 9 Three common ways of creating a pressure difference for sample introduction in capillary electrophoresis: (a) raising the sample vessel above the electrolyte vessel; (b) pressure difference; (c) vacuum created at the end of the capillary tube opposite to the sample introduction

Electromigration. This involves dipping the sampling end of the capillary tube into the sampling vessel while an electric field is applied. This system is simpler than applying a pressure differential in that no extra equipment is needed.

Pressure differentials are used for most analyses because electromigrative injection causes the preferential movement of ions, and hence a bias in sampling. However, electromigration has to be used for capillary gel electrophoresis (CGE) because the gel cannot be put under pressure.

Detection methods

Ultraviolet/visible

The detection techniques used in CE are similar to those used for HPLC – *ie* using ultraviolet/visible light or laser light in conjunction with added fluorophores – molecules that fluoresce in the presence of light of a certain wavelength. This process is called laser induced fluorescence (LIF). The principle of detection is based on Beer's law – *ie* the light absorbed is proportional to the component concentration and the optical path length. (See chapter on Atomic spectrometry). With the conditions used for CE – capillary tubes with diameters as small as 20×10^{-6} m and tiny volumes in the region of 10^{-8} dm^3 – detection appears to be extremely insensitive. The high efficiencies and resolutions achieved by separating through CE at first sight seem to be compromised by the limitations of the detection system. However, CE detection techniques are now moving towards the resolution of single molecules and, in most cases, detection in CE is far more sensitive than in HPLC. In HPLC, the sample is diluted before analysis, whereas in CE the concentration within a sample volume can be increased up to a million times by pre-treatment. Furthermore, the separate zones of the components are very tight and highly concentrated.

The tiny optical path length does pose problems, but this can be partially addressed by modifying the capillary tube in the detecting region and increasing the aperture to let in more light.

RS•C

Ultraviolet/visible detection systems have the instrument electronics continuously comparing the intensity of the reference beam passing through zero sample (I_o) with the intensity passing through the analyte in the sample cell (I_s). Beer's law is followed:

$$\log_{10}(I_o/I_s) = \varepsilon l c$$

where

I_o = the light intensity of the reference beam;

I_s = the light intensity of the beam passing through the sample cell;

ε = the molar absorptivity coefficient for a particular analyte;

l = the path length through the sample cell in metres, m; and

c = the concentration of the analyte in mol dm^{-3}.

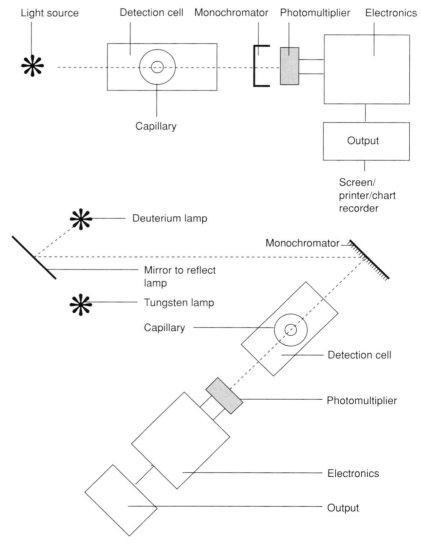

Figure 10 Methods of detection

There are two types of detection systems using ultraviolet/visible fixed wavelength detectors and variable wavelength detectors (Figure 10). The light source for a fixed wavelength detector is usually a zinc, cadmium or mercury vapour lamp and the

RS•C

intensity of the transmitted light is detected by a photodiode or a photomultiplier tube. Fixed wavelength detectors are simple, cheap, robust and sensitive. They are usually set to receive at a wavelength of *ca* 2.3×10^{-7} m, and because ultraviolet spectra are broad, this is adequate for absorbance by component molecules that, for most purposes, are sensitive at this wavelength.

Variable wavelength detectors use two different lamps, usually deuterium and tungsten, the latter for wavelengths of visible light. The position of the mirror determines which light is selected. The monochromator is incorporated into the detector system and the change in wavelengths is made by adjustment of the diffraction grating automatically. Since the chosen wavelength is rarely changed during analysis, the option of variability is not a particular advantage. As the graph shows (Figure 11), fixed wavelength detectors are more sensitive. The lower sensitivity of variable wavelength detectors is mainly due to the low output of the deuterium lamp.

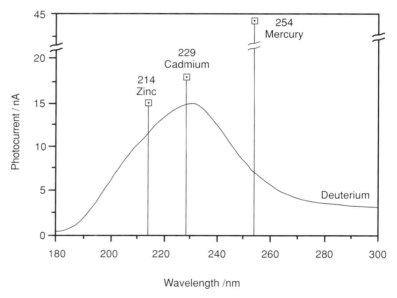

Figure 11 Graph showing the comparison between optical throughput for fixed and variable (deuterium) wavelength detectors

Improving detection sensitivity

Just as the high concentrations of analyte zones enhance the sensitivity of signals, so extending the path length can give greater sensitivity for detection. Two examples used to extend path length are bubble cells (Figure 12) and Z-cells (Figure 13) – so called because of the shape of the capillary tube in the detection area.

A bubble cell is directly fabricated into the capillary tube and is a bubble-shaped swelling along the axis of the capillary tubing. The advantages of the bubble cell are:

■ increasing the length of the light path; and

■ increasing the optical flux – *ie* the amount of light available for detection.

One drawback with bubble cells is that two closely migrating but separate bands may enter the bubble cell together, and both bands may be within the diameter of the light beam of the detector.

RS•C

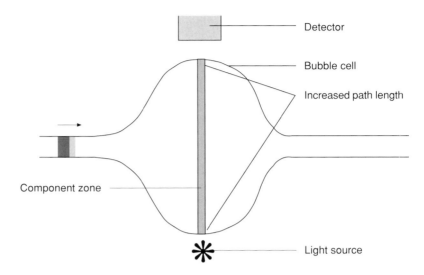

Figure 12 A bubble cell

Another way of enhancing path length is to use a Z-cell incorporated into the capillary tubing at the detector window. A Z-cell can increase the path length considerably but the problems with merging of zones increase.

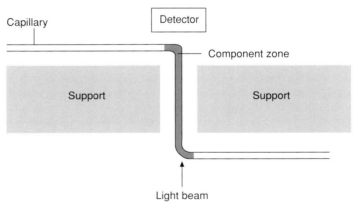

Figure 13 A high sensitivity optical Z-cell

Finally, the simplest way to increase the path length is to increase the aperture to let in more light. However, this does result in a slight loss of resolution.

Laser induced fluorescence (LIF)

In LIF a section of the capillary tube is illuminated by laser light directed from the laser source through a fibre-optic cable. The fluorescence is collected by an ellipsoidal mirror and focused back on to the photomultiplier tube. A beam block cuts out scattered reflected light.

Fluorescence occurs when the laser light interacts with fluorophores added to the sample. The lasers used most commonly are low cost argon-ion, helium-cadmium and helium-neon lasers. These all emit light at wavelengths which correspond to the peak fluorescing wavelengths of the most widely used fluorophores.

RS•C

Laser induced fluorescence detection has a number of advantages over ultraviolet/visible light.

■ Better focusing capabilities (the fibre optic lens can be used inside the capillary tube) allowing the excitation energy to be more effectively applied to a small sample volume.

■ Monochromaticity of the laser light reduces stray light levels.

■ Improved selectivity and sensitivity.

Laser induced fluorescence is particularly suitable for detecting trace amounts of bio-analytes such as DNA fragments. Amino acid analysis is done using LIF because of the relatively small amounts used. The new generation of CE systems and LIF detectors makes it possible to detect single molecules of labelled DNA.

Capillary electrophoresis – mass spectrometry (CE–MS)

Coupling a CE system to a quadrupole mass spectrometer gives a separating system with a large resolving capacity, combined with a very powerful and structurally-specific detector. The sample is sprayed into the mass spectrometer once it has passed the ultraviolet/visible absorbance detector in the capillary tube. This system can be used for analysis of a wide range of specimens, including protein sequences, pesticides and industrial products in the petroleum industry. It can also provide a complete assay for environmental monitoring of river water.

The advantage of CE–MS is that there is a high ratio of molecular ions in relation to ion fragments formed. The main problem in combining mass spectrometry with CE is generating compatible flow rates from the capillary tube to the mass spectrometer. First, the flow rate needs to be increased to produce an effective spray of ions (electrospray) at the interface and, secondly, a medium needs to be provided in which individual ions can be easily produced for the mass spectrometer. The problem is overcome by using a make-up or sheathing liquid which runs in an outer concentric tube in the same direction as the sample in the capillary tube (Figure 14).

RS•C

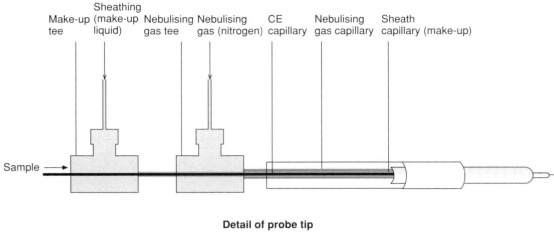

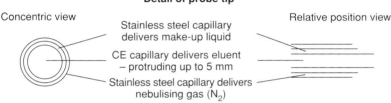

Figure 14 Diagram to show CE/MS interface

The sheathing liquid is usually made up of aqueous mixtures of organic liquids such as methanol, methanoic acid, ethanenitrile and propanone. This liquid meets the effluent specimen at the interface between the CE tube and the mass spectrometer. The higher flow rate of the sheathing liquid enables the electrospray to be done efficiently.

Nitrogen gas runs in the outer tube of the tri-axial array. This gas helps to nebulise the specimen as it enters the mass spectrometer.

CE–MS is a highly efficient combined separating and detecting device. However, there are problems with this system. Most commonly used buffers are non-volatile, making it difficult to vaporise the electrospray. Also mass spectrometry is not compatible with micellar electrokinetic capillary chromatography (MECC) (see page 121) because of interference from the surfactants. In the future, electrospray mass spectrometers that accept much lower liquid flows – *ie* nano-flow systems – may be advantageous for CE–MS applications.

Explaining different modes

Capillary zone electrophoresis (CZE). The medium for CZE is a buffer electrolyte solution plus the sample. This is the simplest mode of CE. It is most commonly used for the detecting small ions and peptides containing typically up to ten amino acids.

Capillary gel electrophoresis (CGE). Capillary gel electrophoresis (Figure 15) is similar to polyacrylamide gel electrophoresis, where large biomolecules are separated by the sieving effect of a gel. Capillary gel electrophoresis has higher separation efficiencies than conventional gel electrophoresis – between 10 million–20 million theoretical plates. Macromolecules have smaller diffusion coefficients in gel-filled media and hence higher separation efficiencies (Box 2, p 114). The advantages of CGE over conventional gel electrophoresis include:

■ the need for only small samples;

■ higher sensitivity;

RS•C

- direct detection – *eg* no need for staining;

- faster analysis time; and

- better resolution.

Capillary gel electrophoresis can resolve down to single base pairs of DNA fragments.

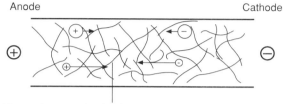

Anode   Cathode

Obstructing strands of gel cause large ions to move more slowly

Figure 15 Schematic diagram of CGE

Capillary gel electrophoresis (CGE) is different from CZE in that EOF does not occur because of the presence of the gel. Using coatings on the walls of the capillary tubes helps to eliminate EOF (Box 1, p.113).

The polyacrylamide or agarose gels that are used for CGE are prepared within the capillary tubing. The gels can be prepared either with or without crosslinkages, though it is essential to provide linkages to the capillary tube wall. The gels are prepared with bifunctional reagents such as 3-methacryloxypropyltrimethoxysilane or 3-methacryloxypropyldimethyoxysilane. These reagents have a functional group at one end that can bind chemically to silanol groups (Si–OH) on the capillary tube wall. At the opposite end there is another reactive group which forms a covalent bond with the polymeric gel.

Soluble polymers such as hydroxyethylcellulose can be dissolved in buffer solutions for applications that do not need high resolution. One limitation of CGE is that gel columns are sometimes vulnerable to bubble formation and contamination.

Micellar electrokinetic capillary chromatography (MECC). Micellar electrokinetic capillary chromatography is different from other CE separation techniques in that it is capable of separating and resolving uncharged molecules as well as ions. The technique relies on the establishing a partition coefficient between a micellar phase and a buffer solution phase.

A micelle (Figure 16) is formed when a surfactant is added to water at a concentration above a value known as its critical micelle concentration (cmc). Surfactants have both a hydrophobic and a hydrophilic character – *ie* the molecules consist of polar 'head' groups and non-polar hydrocarbon 'tails'. When the concentration of surfactants exceeds the cmc they self-aggregate within the solution with the hydrophobic tails pointing inwards away from the aqueous solution.

RS•C

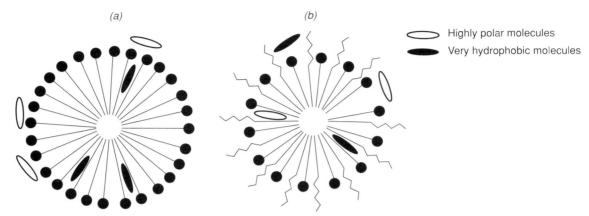

Figure 16 A schematic representation of the interaction between the analyte and the (a) ionic or (b) mixed micelle of ionic and non-ionic surfactants

The most commonly used surfactant is sodium dodecyl sulfate $(CH_3(CH_2)_{10}CH_2OSO_3^-Na^+, SDS)$. Since it has a negatively-charged head it is known as an anionic surfactant. This can be thought of as small droplets of oil with densely negatively charged surfaces. Very hydrophobic molecules such as the dye Sudan III dissolve completely inside the micelle with very few, if any, molecules remaining in the bulk solution (Figure 16a). However, highly polar molecules like methanoic acid do not dissolve inside the micelle and remain in the bulk solution. Neutral molecules are thus partitioned between the micelles and the bulk solution depending on their hydrophobicity – *ie* their affinity for non-polar substances such as the hydrocarbon tails of the micelles.

The individual migration times of the neutral molecules depend on their interactions within the micelle. For example, with a SDS surfactant the micelle tends to migrate towards the anode because it is negatively charged. However, when the resultant EOF is greater than the electrophoretic flow all species are flushed in the direction of the cathode, depending on their charge and the hydrophobicity of the molecules. Anions tend to have the shortest migration time because of repulsion from the negative micelles, with the cations migrating last as a result of attraction to the micelles. The molecules move at a speed that depends primarily on their hydrophobicity. Very hydrophobic molecules (being distributed almost entirely within the micelles) move slowest and hydrophilic molecules move at the speed of the resultant EOF in the bulk solution. Neutral molecules thus move in bands that are defined by the extent of their distribution between the micelles and the bulk solution (Figure 17).

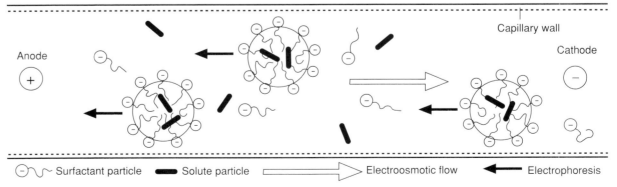

Figure 17 A schematic representation of the separation principle of MECC. The micelles migrate towards the cathode because the EOF is greater than the electrophoretic flow

RS•C

Capillary isoelectric focusing (CIEF)

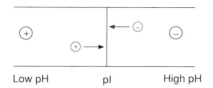

pI = pH corresponding to the isoelectric point of an amino acid, peptide or protein. (See section on Gel electrophoresis).

Figure 18 Isoelectric focusing

The principle of CIEF (Figure 18) is similar to that of isoelectric focusing in slab or column gel electrophoresis. However, in CIEF the electrophoresis can be done in free solution. The sample is introduced with buffers which have different pH values from each other, forming a continuum or gradient between two pH values such as pH 4 and pH 8. It is then possible for the component molecules in the sample to separate out according to their isoelectric points (pI). Small amounts of additives, such as methyl cellulose or hydroxypropyl methyl cellulose, are added to the sample or the buffer, allowing CIEF to be performed in untreated silica capillary tubes.

In gel isoelectric focusing, the components are resolved into separate bands that can then be stained. However, in CIEF the problem is that the components are focused into separate stationary bands that then have to migrate past a detector window. The bands are mobilised by simply raising one end of the capillary tube and allowing the sample to flow in the direction of the detector cell (Figure 19).

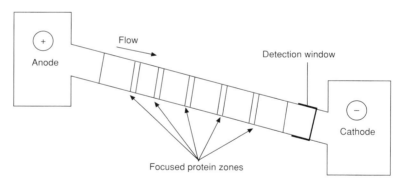

Figure 19 Samples are focused and then driven towards the detector by low pressure

The resolving power of CIEF is expressed as the difference in pI of the two species being separated. With immobilised pH gradients (see chapter on Gel electrophoresis) this can be as low as 0.001 pI units. The efficiency – *ie* the number of theoretical plates – depends on the shallow rate of change of pH with capillary tube distance (a shallow pH gradient), a high electric field and efficient cooling of the capillary tube.

Chiral recognition

Separating the enantiomers of chiral compounds has become increasingly important in pharmaceutical, agricultural and forensic applications. Regulatory pressures and the potential cost of leaving non-effective or toxic enantiomers in racemic mixtures have made it necessary to look at ways of resolving these mixtures. A quick and simple method is required for measuring optical purity and for separating enantiomers of drugs such as antileprosy drugs and steroids.

RS•C

Enantiomers can be differentiated using compounds such as cyclodextrins that interact stereoselectively with the enantiomers. Cyclodextrins (CDs) are oligosaccharides with a truncated cylindrical shape, and inclusion complexes can be formed in the molecular cavity. The outside surfaces of CDs are hydrophilic, encouraging dissolution in the bulk solvent, whereas the interior is hydrophobic.

The secondary hydroxyl rim of the CD consists of a number of chiral centres that interact selectively with the guest enantiomers. Some modified CDs are more soluble in water – eg sulfonated forms – and have a range of enantiomeric selectivities. S-enantiomers tend to form more stable complexes than R-isomers, and have slightly longer migration times.

Comparing CE with other separation techniques

Capillary electrophoresis is developing quickly and some of the present limitations may be overcome by future instruments. So any comparisons have to anticipate future possibilities. There are a number of techniques that perform similar – and sometimes complementary functions to CE. The question is, which technique provides a better solution to a problem in terms of accuracy, precision, speed and cost.

Comparison with high performance liquid chromatography (HPLC)

The most significant factor when comparing CE with HPLC is the difference in flow profile, leading to a greater number of theoretical plates in CE. The liquid flow under pressure through a packed chromatography column has a parabolic leading profile resulting in band spreading. The CE leading profile is flat, which reduces band spreading and increases efficiency (see p. 113).

Capillary electrophoresis	High performance liquid chromatography
The components in a sample are separated through differential migration rates in an applied electric field	The components in a sample are separated through differential partitioning of solute between a stationary phase and a moving phase
A flat (plug-flow) leading profile leads to a greater number of theoretical plates	A parabolic leading profile
Flow rate: 10^{-6} dm^3 min^{-1}	Flow rate: 10^{-3} dm^3 min^{-1}
Injection volume: 10^{-9} dm^3	Injection volume: 10^{-6} dm^3
Detection cells are situated on the column.	Detection cells are adjacent to the column

Table 1 The differences between CE and HPLC

Detection is simpler in CE because the detection cells are on-column, but it is also harder because of the tiny volumes. Capillary electrophoresis detection systems have been derived from those used for HPLC, but work is now being done to develop detection systems, such as conductivity cells, that are specifically designed for CE.

High performance liquid chromatography and capillary electrophoresis can produce complementary data sets and can be interfaced to form a two-dimensional separating system. Since the separation principles are different, the information gathered from one technique can be used to enhance the other. For example, the enzyme ribonuclease B is digested with chymotrypsin – selectively breaking down the enzyme – producing fragments that enables amino acid sequencing. The fragments are collected then separated using HPLC. A fragment (one peak) is collected from the HPLC column and is then analysed using CE, resolving into three smaller fragments, each one of which is collected and sequenced with initial yields in the picomole (10^{-12}) range.

RS•C

Electrokinetic capillary chromatography   **Box 3**

This is a combination of CE and HPLC. A capillary tube is filled with a packing material, the sample and buffer are introduced and a voltage is applied across the ends of the capillary tube to produce an EOF. This removes the need to pump fluid through the capillary tube, and more importantly, produces highly efficient plug-flow, giving high efficiencies. The components are separated and retained on the stationary phase depending on the strength of the interactions between the liquid sample and the packing material.

Comparison with gel electrophoresis

Compared with CE, only relatively low voltages can be used in slab gel electrophoresis due to heating effects. In contrast, high voltages can be used in CE because of the rate at which heat is dissipated, and the very small volumes of liquid involved. In addition, CE is easier to automate, is much faster and requires much smaller samples. While an apparent advantage of conventional gel electrophoresis is that it can be used for preparation and purification, the possibilities of using capillary arrays could challenge this advantage. However, gel electrophoresis has been used to characterise and speciate materials, and many standard methods have been established.

Applications of CE

Haemoglobin analysis

Abnormalities in the structure of haemoglobin are associated with a variety of congenital diseases. These include sickle-cell anaemia, thalassaemia syndromes and conditions associated with diabetes. Diagnosis of these haemoglobin variants is clinically important in treating these disorders.

There are several varieties of haemoglobin molecules. These are polypeptide tetramers consisting of two pairs of slightly different peptide ('globin') chains, each chain being linked to a planar haem group (Figure 20).

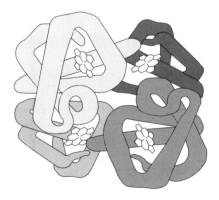

Figure 20 The structure of haemoglobin showing the arrangement of alpha and beta chains

Data from the US show that the predominant normal human haemoglobin for over 95% of the population, is HbA ($\alpha 2 \beta 2$). Determining the presence of abnormal haemoglobins such as those in sickle cell disease is important for its diagnosis, and as a follow-up to blood transfusions. The red blood cells of people afflicted with sickle-cell anaemia contain an abnormal form of haemoglobin, known as HbS. A glutamic acid ($HO_2CCH_2CH_2CH(NH_2)CO_2H$) residue in the β-chain of the normal HbA is replaced by an uncharged valine ($(CH_3)_2CHCH(NH_2)CO_2H$) residue. This affects

RS•C

solubility and the haemoglobin tends to precipitate causing the red blood cells to 'sickle' (*ie* form a curve) and burst.

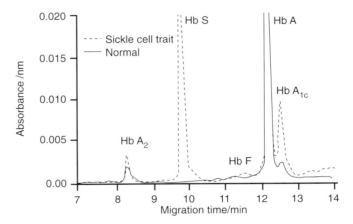

Figure 21 Capillary electropherogram of some haemoglobin constituents in normal blood and in blood from a patient with sickle cell trait. Note that HbS forms a distinct peak well separated from normal Hb

Capillary isoelectric focusing can be used for the rapid and routine analysis of haemoglobin variants, with a resolution as low as 0.02 pI units between variants. Because the EOF needs to be inhibited, the capillary tubes are coated and the analysis is done on samples that form a pH gradient in the capillary tube. After focusing, the components are eluted under pressure at high voltage.

Detecting chemical weapon products

Highly sensitive analytical techniques need to be used for detecting products banned by international conventions. These include chemical weapons excluded by the Chemical Weapons Convention. Visiting international observers need instruments that can provide data quickly and detect small quantities of materials that indicate chemical weapon degradation.

Gas chromatography (GC) is effective for detecting volatile samples, but other methods are needed for non-volatile and aqueous samples which are usually obtained by swabs from surfaces, soils, greases, scrapes of paint and water effluents. Agents such as sarin used in the 1995 Tokyo underground poisonings, degrade to form chemical signature compounds – alkylphosphonic acids, alkylphosphonthioic acids and associated esters. These must be detected in tiny concentrations.

CZE and MECC are reliable identifiers of degradation products over a wide range of chemical weapon agents. Given the tiny amounts of sample often available, this type of investigation is likely to prove challenging but it is accessible through CE.

Separating milk proteins

Food companies need to separate and identify their milk products in different formulations such as ice creams and dried milks. This is particularly important for products that may be stored anywhere from the Arctic to the Equator. Capillary zone electrophoresis can separate a wide variety of proteins in milks (Figure 22).

RS•C

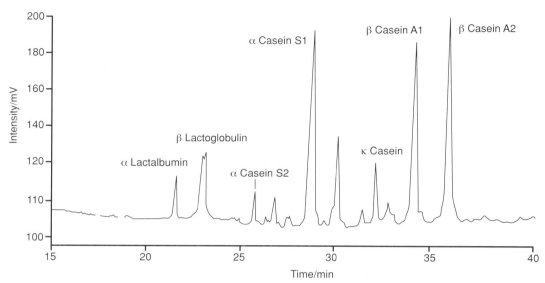

Figure 22 The separation of milk proteins

Gunshot residues

Forensic evidence of a firearm crime usually involves analysing gunshot residues which gives information on the type of weapon used. Gunshot residues are formed from primers consisting of a variety of metal ions including lead, barium, antimony, a propellant of nitrocellulose and nitroglycerine, and a plasticiser. Atomic absorption spectrometry (AAS), amongst other techniques, is used to detect the primer residues, but the chemical composition of primers is being changed in favour of organic constituents – due partly to environmental concerns over heavy metals!

Capillary electrophoresis has been used to detect organic primer residues on the hands of people who have fired guns. Components can be identified in small amounts on the hands of weapon users. Analysing organic residues using CE showed that these organic residues were found only on the hands of the weapon users and not in the general population. However, heavy metal residues were detected using AAS on both the hands of weapon users and in the general population, thereby giving less reliable data.

Detecting drugs

Capillary electrophoresis has been used by scientists at the University of Verona, Italy, to detect illicit drug use by examining hair samples. Hair has a number of advantages in drug detection. It concentrates the drug, and no further metabolism takes place once the drug has passed into the hair. The drug remains fixed in the hair so it can be detected some months after use. Also, suspects experience less distress when some hairs are removed for analysis rather than taking a blood sample. However, bald suspects present an impressive analytical challenge.

Both CZE and MECC are used for these analyses. The techniques complement each other because the analyte is used in the same instrument and only the buffer needs changing, thereby providing a confirmation of results.

Other applications

These include analysing caffeine in tea, coffee and soft drinks. Capillary electrophoresis also has the potential for quick screening in clinical diagnosis. For

RS•C

example, up to 36 different ions can be detected in less than 2 mins when screening the metabolic products in human urine.

In the detergent industry, the active ingredients in washing powders are formulated with components such as whiteners, surfactants, enzymes and anticaking agents. Capillary electrophoresis can quickly resolve the separate components in this complex matrix.

The low detection limits of some CE systems means that these are used to monitor the air in factories involved in the semiconductor industry, and can be used to identify low levels of rust in samples from the water-cooling systems of nuclear reactors.

Some specific applications include:

- separating proteins found in human tears;

- using MECC for separating the water-soluble B vitamins;

- using CZE for separating ascorbic acid isomers in vitamin C;

- separating rare earth metals; and

- analysing genetically modified insulin, organic acids in wine, and spider venom.

Future developments

Capillary electrophoresis on a chip. A coiled capillary tube is etched into a glass chip about the size of a postage stamp. The voltage supply comes from a small battery so it can be used in the 'field'. The sample components are separated in the chip. These chips have potential in clinical and forensic science. For example, different chemicals might be sampled in a patient's serum by inserting chips in the arm. One type of buffer will be run through one chip, say, to separate salts. Another buffer in another chip will be used for protein analysis. Yet another might detect sugars. Forensic applications including chemical weapons testing and gunshot residues are other possibilities.

Capillary arrays. Many capillary tubes can be used simultaneously for analysing the same sample – *eg* serum profiling. This can give a huge throughput providing reliable results that can be cross-checked. At the same time the array can contain capillary tubes working in different modes, some for CZE, others for MECC, CGE and CIEF all giving optimum results and different pieces of information depending on the components being separated.

Data interpretation and problem solving

Problem
Six peptides need to be separated from a mixture. All the peptides have at least one amino group that can be protonated. All the peptides are water soluble to varying degrees, but some have more polar amino acid residues than others.

The sample is prepared in water, the concentrations of the peptides varying between 6–25 $\mu g\ cm^{-3}$.

Three buffers are available: borate at pH 9, phosphate at a low pH, and a SDS-borate mixture.

Using CZE in a conventional silica capillary tube, suggest what you might expect to happen when you use each buffer. Explain your choice and what each electropherogram might look like.

RS•C

Three electropherograms, Figure 23 (a), (b) and (c) are shown of what actually did happen when we ran these samples using CE. Identify the electropherogram with each buffer system. (It is not possible to identify the individual peptides. This is done by running standards of the pure peptides to establish their migration times).

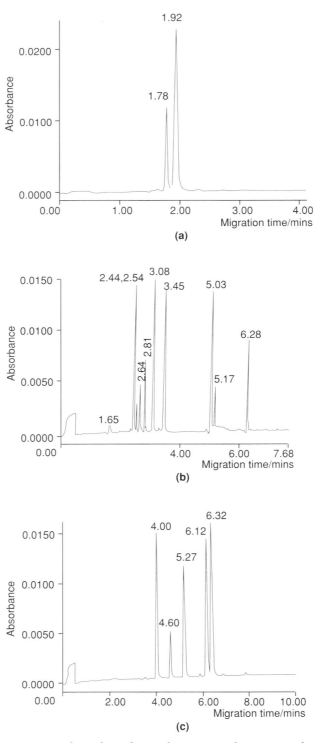

Figure 23 (a) , (b) and (c) Electropherograms of SDSB, P and B buffers
(not in that order)

RS•C

Solution

All the peptides have an ionisable amino and carboxyl group. Suppose a low pH buffer is used, such as phosphate. It is reasonable to assume that all the amino groups are protonated, so that the overall charge on each peptide is positive. Electrophoretic flow is from the anode towards the cathode and the peptides separate according to their mass to charge ratio (m/z) . At a low pH the EOF is relatively small and is unlikely to disturb the movement of the peptides. There may be some problems in separating the peptides with less polar components.

At a high pH using a borate buffer, the amino groups are not ionised and the carboxyl groups are only slightly ionised, as the -COOH group is only weakly acidic. While some separation is expected on the basis of the mass to charge ratios (m/z), the EOF is considerable but opposes the flow of the peptides. This is not necessarily a problem, but some of the peptides may not have a large enough mass to charge (m/z) ratio to enable them to move towards the anode, whereas others would. However, given the weak charge, it is unlikely that there is sufficient resolution between some peptides.

Using a SDS-borate buffer (that is a MECC separation) could overcome the problem of non-separation of non-polar peptides, whereas an anionic micelle could help to resolve the non-polar entities.

Figure (a) is an electropherogram of the separation achieved using borate, (b) is SDS-borate and (c) is phosphate. The separation through borate is relatively weak and it is likely that the strongly polar peptides are best resolved. The SDS-borate seems to have achieved the best separation of the peptide components. The migration times of the peptides are influenced by the buffer system.

RS•C

X-ray crystallography

Introduction

There are many analytical techniques that give both qualitative and quantitative information but there are only a few that produce structural information. X-ray crystallography (XRC) is the most well-known of these techniques. When X-rays scattered by a crystal are detected – either on film or electronically – they form a pattern. Interpreting this pattern helps to explain the crystal structure of a sample (Figure 1). For example, in 1938 Dorothy Crowfoot Hodgkin confirmed that benzene rings are planar using XRC. Hodgkin also discovered the structures of penicillin and vitamin B_{12} using this technique.

Figure 1 Example of a diffractogram pattern
(Reproduced by permission of the EPSRC National X-ray crystallography service.)

Theory

Before exploring how XRC provides structural information, it is necessary to look at the general structures of crystals.

Crystals
Crystals are usually solids, recognised by their repeating geometric pattern. They have a regular internal or lattice structure – a three-dimensional repeating pattern of atoms or ions. Solids that do not have a crystal structure are called amorphous.

The unit cell
The smallest three-dimensional repeating unit within a crystal is called the unit cell (Figure 2). Unit cell parameters in the form of angles and length are used to define the size of the unit cell. With a knowledge of the dimensions of the unit cell a diffraction pattern can be predicted.

RS•C

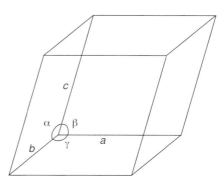

Figure 2 Diagram of a typical unit cell showing angles and sides

Close packing   **Box 1**

The structures of most unit cells can be explained by the simple packing of small hard spheres. The arrangement of the spheres in Figure 3 is close packed. A three-dimensional close packed arrangement is made by adding a second layer of spheres on top of the first layer (Figure 4).

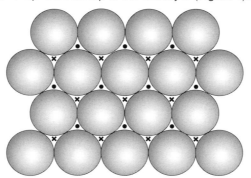

Figure 3 Close packed spheres. Note the shape of the spacing between the spheres
(Adapted with permission from L. Smart and E. Moore, *Solid State Chemistry,* London: Chapman & Hall, 1992.)

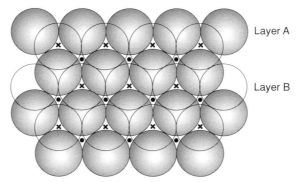

Figure 4 Diagram to show arrangement of layers
(Adapted with permission from L. Smart and E. Moore, *Solid State Chemistry,* London: Chapman & Hall, 1992.)

There are two possible ways of arranging a third layer.

1. The third layer of spheres is positioned directly over the first layer. This is an ABABA .... arrangement. This is known as hexagonal close packing (hcp) (Figure 5).

2. The third layer is placed over the remaining spaces between the spheres indicated by • but its position does not correspond to the other two producing an ABCABC... arrangement. This is known as cubic close packing (ccp) (Figure 5).

RS•C

Close packing

Box 1 continued

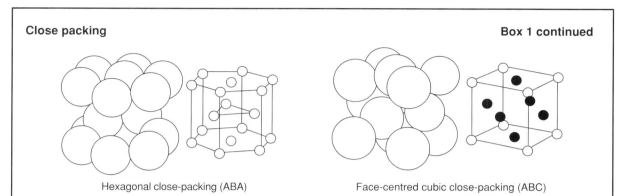

Hexagonal close-packing (ABA) Face-centred cubic close-packing (ABC)

Figure 5 Diagrams to show hcp and ccp

Only some metals (Figure 6) and noble gases (when they have crystallised as a solid at low temperatures) have close packed systems. The atoms of metals and noble gases can be considered as hard perfect spheres of equal size. These 'ideal' systems are uncomplicated by distortions which arise from covalencies between atoms and varying electrostatic forces caused by ionic or polar effects. However, even the unit cell structures of relatively complex molecules can be derived from these structures.

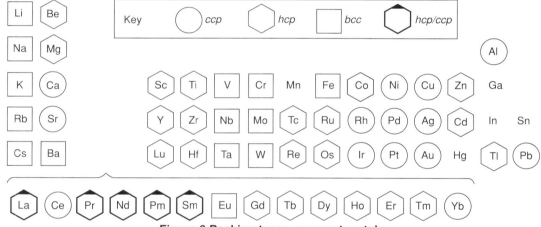

Figure 6 Packing types amongst metals

(Adapted with permission from L. Smart and E. Moore, *Solid State Chemistry,* London: Chapman & Hall, 1992.)

For example, copper shows that close packing is related to the unit cell structure. A single sphere in either of the close-packed structures has 12 equidistant neighbours in three-dimensions. The unit cell for copper is a face-centred cube (fcc) (*Figure 7a*), where each of the atoms in the unit cell has 12 nearest neighbours – all equidistant. Not all metals are close packed. Iron, below its transition temperature, has a body centred cubic structure (bcc) (*Figure 7b*), which is less efficiently packed than the close packed structure. Hence each iron atom is surrounded by only eight equidistant neighbours. Figure 7 shows the structures of both copper and iron.

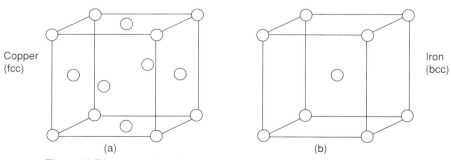

Copper (fcc)   Iron (bcc)

(a)   (b)

Figure 7 Diagram showing structures of copper (a) and iron (b)

RS•C

Types of unit cell
There are various types of unit cell.

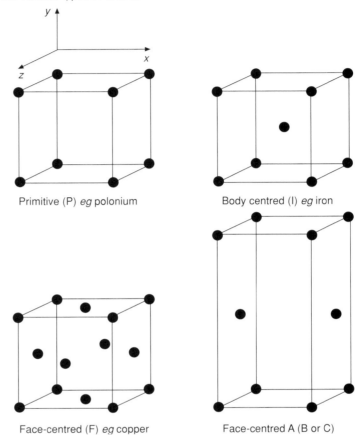

Primitive (P) *eg* polonium

Body centred (I) *eg* iron

Face-centred (F) *eg* copper

Face-centred A (B or C)

Figure 8 Types of unit cell
(Adapted with permission from L. Smart and E. Moore, *Solid State Chemistry*, London: Chapman & Hall, 1992.)

Primitive cube. In this unit cell structure single atoms, ions or molecules are located at each of the corners of a cube. Polonium is the only metal that has this structure.

Body centred cube (bcc). Atoms or ions are located at the corners of the cube and at the centre of the body diagonals – *eg* iron.

Face centred cube (fcc). Atoms or ions are at the corners and intersections of the face diagonals – *eg* copper.

The face-centred unit (symbol A, B or C) has the atoms or ions at the corners and one atom or ion in the centre of one pair of opposite faces *eg* an A-centred cell has atoms or ions in the centres of the bc faces.

RS•C

Crystal systems

Seven crystal systems are defined by the relative lengths of the three sides of a unit cell and the interfacial angles as shown in Figure 9.

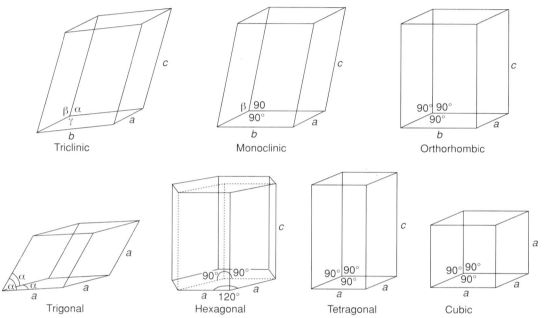

Figure 9 Seven crystal systems

(Adapted with permission from L. Smart and E. Moore, *Solid State Chemistry*, London: Chapman & Hall, 1992.)

These seven crystal systems together with the four types of unit cell give 14 varieties of unit cell called the Bravais lattices (Figure 10).

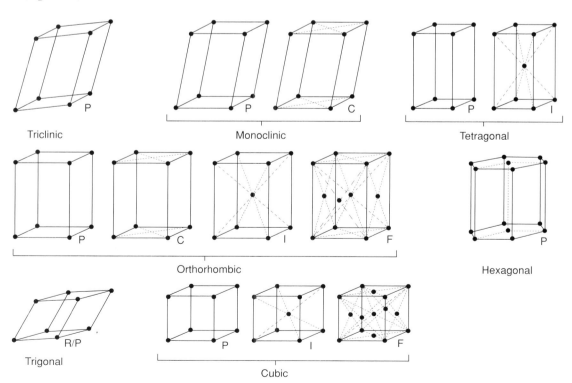

Figure 10 The 14 Bravais lattices

(Adapted with permission from L. Smart and E. Moore, *Solid State Chemistry* London: Chapman & Hall, 1992.)

RS•C

Families of planes

Crystals tend to grow with their faces parallel to the faces of the unit cell, and to other planes in the crystal which have a high density of atoms, ions or molecules. Splitting occurs at the sites of these planes. For example, in the primitive cube, parallel planes can be drawn which include all the particles of a unit cell.

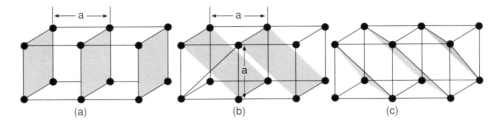

Figure 11 Diagram showing examples of planes in a primitive unit cell

These planes of particles are responsible for the diffraction of X-rays and the resulting patterns. The categorisation of planes in a crystal – known as indexing – is essential for analysing the diffraction pattern.

Indexing crystal planes

The term Miller indices describes the positions and spacings of crystal planes. Particles are considered as points at the interceptions of the axes of a cube. There are two pieces of information which help to define the Miller indices of a unit cell.

1. The axes of the unit cell have the coordinates x, y, z. (Figure 11)

2. The number of spacings, based on unit cell dimensions, between the planes in the:
   x-direction is h;
   y-direction is k and
   z-direction is l (Figure 12).

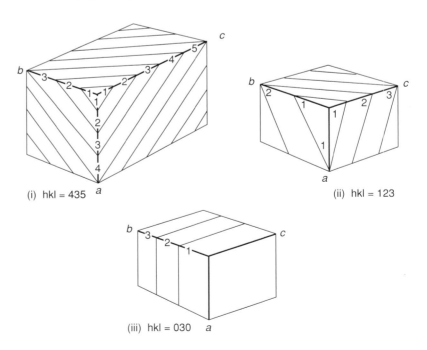

Figure 12 Diagrams to illustrate h, k, l indices

RS•C

Miller indices for the primitive simple cube

Referring to Figure 11, the black circles define the lattice points. There are three different sets of planes which account for the lattice points. The cube has side a. Two unit cells are drawn showing three planes.

The Miller indices are defined as 100. The spacing is $1:\infty:\infty$. The Miller index is the reciprocal of the spacing – ie $1/1:1/\infty:1\infty/$ or $1:0:0$.

The separation between the planes is symbolised as d_{100}, and equals a for primitive systems.

Figures 13 and 14 show planes in the fcc and bcc systems. Figure 14(b) shows a set of planes in the fcc cell. The Miller indices are 110. The perpendicular distance, d_{110}, is equal to the length of half the diagonal of the square face.

From Pythagoras' theorem,

$$2 \times d_{110}^2 = a^2$$
$$d_{110}^2 = a^2/2$$

$$d_{110} = \sqrt{(a^2/2)} = a/\sqrt{2}$$

or

$$d_{110} = a/\sqrt{1^2 + 1^2 + 0^2} \quad \text{or } d_{110} = a/(1^2 + 1^2 + 0^2)^{1/2}$$

Figure 14(c) shows a third set of planes that can be drawn between the lattice points.

The Miller indices are 111.

The separation between the planes can similarly be shown to be:

$$d_{111} = a/\sqrt{3}$$

or

$$d_{111} = a/\sqrt{1^2 + 1^2 + 1^2} \quad \text{or } d_{111} = a/(1^2 + 1^2 + 1^2)^{1/2}$$

A single general equation can be used to define the spacings of the three sets of planes.

$$d_{hkl} = a/(h^2 + k^2 + l^2)^{1/2}$$

This equation holds for any spacings in the cubic system. Further examples are given in Figure 15.

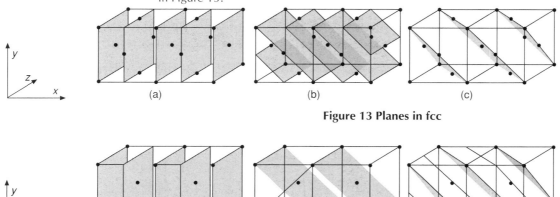

(a) (b) (c)

Figure 13 Planes in fcc

(a) (b) (c)

Figure 14 Planes in bcc

RS•C

Primitive	$d_{100} = a$	$d_{110} = \frac{a}{\sqrt{2}}$	$d_{111} = \frac{a}{\sqrt{3}}$
Face-centered	$d_{200} = \frac{a}{2}$	$d_{220} = \frac{a}{2\sqrt{2}}$	$d_{111} = \frac{a}{\sqrt{3}}$
Body-centered	$d_{200} = \frac{a}{2}$	$d_{110} = \frac{a}{\sqrt{2}}$	$d_{222} = \frac{a}{2\sqrt{3}}$

Figure 15 Interplanar spacing

Knowledge of these differences in spacings is helpful in identifying the type of unit cell from diffraction data.

The meaning of diffraction patterns

The unit cell is an imaginary geometric figure described by the positions of particles – ie atoms or ions – in a repeating unit. It was this concept that led the scientist von Laue to use a crystal as a three-dimensional diffraction grating. He placed a copper(II) sulfate crystal between an X-ray source and a photographic plate. A pattern of spots appeared on the plate as the crystal diffracted the X-ray beams. W.H. Bragg and his son Lawrence Bragg consolidated Laue's observations, producing clear diffraction patterns to determine many different structures.

There are two features of a good diffraction pattern such as those obtained by the Braggs.

■ The spacings between the spots provide information about the size and shape of the unit cell.

■ The spots have different intensities. Information about the arrangement of the particles can be derived from the intensities. The intensity increases with the atomic number of the atoms.

Interpreting the pattern on the photographic plate gives information about the internal structure of the crystal. The interaction between X-rays and the planes of particles within the crystal is responsible for the pattern. X-rays are diffracted by the atom arrays along the crystal planes.

Diffraction

Diffraction occurs when light waves meet a barrier – eg a diffraction grating – with one or more slits of similar magnitude to the wavelength. After passing through the barrier the light spreads out into that region normally expected to be in shadow. The explanation for diffraction is that each point on the wavefront within the slit acts as a point source, radiating in all directions. Regions of light or dark bands appear on a screen due to the interference of these secondary waves (Figure 16).

RS•C

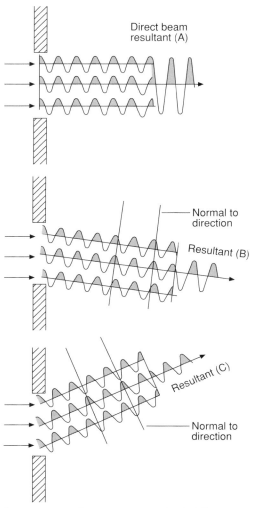

Figure 16 Interference as a result of diffraction

The breadth of each band in the diffraction pattern is related to the width of the slit. The narrower the slit the broader the bands (Figure 17).

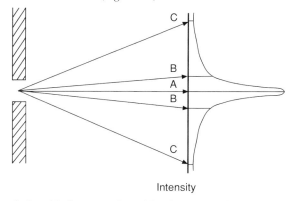

Figure 17 Relationship between breadth of bands and slits

In crystals the spaces between the planes of particles act as slits but are too small to diffract visible light. However, they can diffract shorter wavelength radiation, such as X-rays with wavelengths of about 10^{-10}m. The diffracting structures are the electron

RS•C

clouds of atoms in the crystals rather than slits. The electron clouds interact with the electric field of the X-rays. The regularly repeating planes of atoms in the crystal produce interference. When the beams diffracted from one unit cell are in phase with the beams from adjacent unit cells, constructive interference occurs producing an intense diffracting beam. The intensity of the diffracting beam depends on the wavelength of the monochromatic – *ie* 'pure' or of one wavelength – X-rays, but also increases with the atomic numbers of the atoms. Hydrogen, with only one electron, has no detectable effect on X-rays, and the positions of hydrogen atoms have to be inferred.

The interaction of X-rays with crystal planes – the Bragg equation

The diffraction of X-rays by a crystal can be quantified. Imagine atoms aligned along layers of planes in a crystal. These layers diffract X-rays. There are two possibilities when X-rays are incident on successive planes of the crystal at a specific distance (*d*) apart – the X-ray waves can reinforce (Figure 18), or they can cancel out.

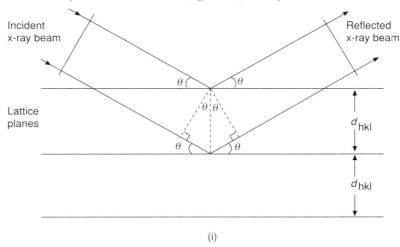

(i)

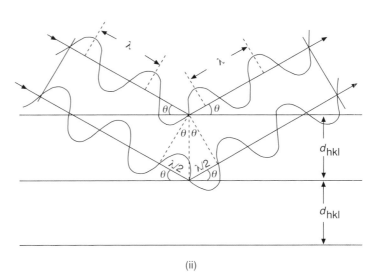

(ii)

Figure 18 Reinforcement of reflected X-rays
(Adapted with permission from C. Whiston, X-ray Methods,
Chichester: John Wiley & Sons on behalf of ACOL, 1987.)

RS•C

The Bragg equation is:

$$n\lambda = 2d\sin\theta$$

where
n = the order of the reflection and is equal to an integral value (1, 2, 3 etc. ...)
λ = the wavelength of the X-ray
d = the interplanar spacing
θ = the angle of incidence with the diffracting plane

A diffraction pattern occurs only when there is constructive interference between the waves incident on the two successive planes. This happens when the path difference between the X-rays is equal to a whole number of wavelengths. Each wave diffraction is known as a 'reflection'. A first order reflection occurs when there is a path difference of one wavelength, a second order reflection when there is a path difference of two wavelengths, and so on.

Obtaining diffraction data from a single crystal
To obtain a diffraction pattern from a single crystal an X-ray diffractometer is used.

There are three main components in the instrument. These are:

- a source of X-rays, usually an X-ray tube;

- a sample holder; and

- a means of detecting the diffracted X-rays. This can be a camera and film – particularly for protein crystals. However, modern computerised diffractometers do not use film. Modern instruments are now used for all types of X-ray structure determination.

Generating the X-rays
X-rays are generated when fast-moving electrons are 'stopped' by a heavy metal target. All X-ray tubes have the following in common:

- a cathode to provide a source of electrons;

- a means of producing a potential difference across the tube to accelerate the electrons;

- a metal target to stop the electrons and produce X-rays – the anode;

- a means of directing the X-rays towards the sample; and

- a means of cooling the tube.

Using the Bragg equation with a knowledge of λ and θ the interplanar spacing (d) can be calculated. Monochromatic radiation is used, produced by absorption through a metal filter.

RS•C

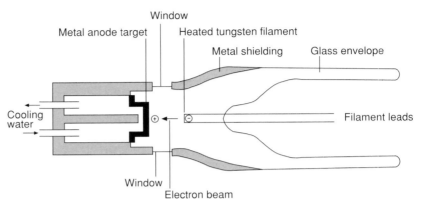

Figure 19 An X-ray tube

Electrons are produced by electrically heating a metal filament acting as a cathode. A large potential difference of around 60 kV, mediated by a high quality transformer and rectifier, accelerates the electrons towards the metal anode target. The tube is evacuated to prevent electrons being impeded in their flight across the tube. Modern X-ray tubes are manufactured in a sealed unit with an existing vacuum. An electrostatic focusing system directs the electrons towards a point on the metal target. When the electrons strike, X-rays radiate from the target. For safety the tube is lead-cladded, and in laboratories using X-ray equipment operators are required to wear badges which monitor exposure. X-rays leave the tube through windows made of a light metal such as beryllium or aluminium.

On impact, only about one per cent of the electron's energy is converted to X-rays, the rest is lost as heat which could result in melting of the target and over-heating of the tube. The target is usually composed of metals such as molybdenum, rhodium, cobalt, tungsten or copper, ie, metals with high melting points and good thermal conductivity. They are cooled by flowing water. Automatic cut-off systems are incorporated into the device in case the cooling system fails.

'White' X-rays and characteristic lines are produced meaning that the radiation consists of a continuum of wavelengths, so metal filters (Box 2) are used to ensure that only a specific wavelength is used for diffraction.

Metal filters Box 2

When electrons strike the metal target two possible interactions occur. An electron may be stopped completely on impact and all its kinetic energy converted into radiation of high energy short wavelength X-rays. Some electrons, however, will lose only part of their energy on striking the metal and will continue to travel and interact with successive atoms, losing progressively more of their kinetic energy, thereby producing a continuous range of wavelengths.

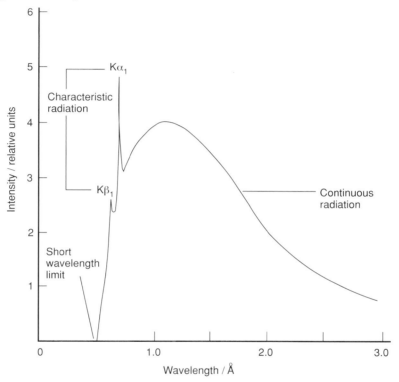

Figure 20 The 25 kV X-ray spectrum of molybdenum

'White' radiation is unsuitable for diffractometry (the Bragg Equation relates to wavelengths of a single value) but monochromatic radiation can be produced by utilising the discontinuities in the spectrum. Sharp peaks are produced, designated $K\alpha_1$, $K\beta_1$ and $L\alpha_1$, etc. These occur when the energy of impact is such that electrons are removed from one of the inner shells and the atom becomes ionised. An electron falls from an outer shell to replace the lost electron, releasing X-rays of a specific wavelength at the same time. (These transitions are discussed in more detail in the section on X-Ray Fluorescence.)

RS•C

Box 2 continued

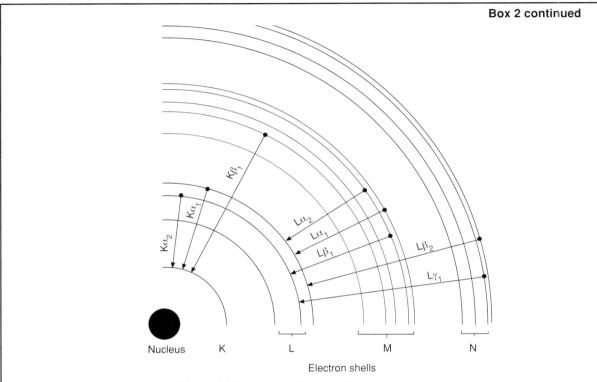

Figure 21 How some K and L lines originate

The wavelengths represented by these sharp peaks are intense and almost monochromatic. A device that can cut out all or most of the 'white' radiation is needed, selecting only one of the $K\alpha_1$, $K\alpha_2$ or $L\alpha$ wavelengths, usually $K\alpha_1$. Filters of another metal which absorb a large part of the radiation can achieve this. As shown in Figure 22, a nickel filter absorbs most of the radiation emitted by a copper target except for the $K\alpha_1$ and a small proportion of the white radiation.

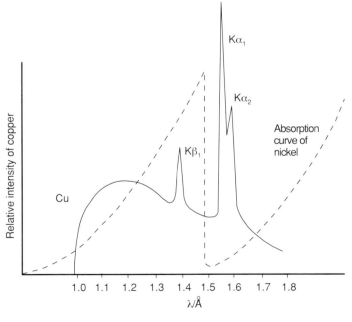

Figure 22 Nickel foil acting as a filter for copper X-rays

RS•C

Variations in the X-ray generating system exist. Many generators incorporate a rotating anode so that a greater DC potential can be applied without overheating. This helps to produce greater X-ray intensity, which is particularly important for weakly diffracting crystals such as those of proteins and other biopolymers. Some modern X-ray tubes are insulated by lightweight ceramic materials.

Those X-rays which pass through the sample undiffracted are absorbed by a beam stop.

The goniometer

The goniometer is a device which orients the crystal in relation to the detector and the incident X-ray beam by rotating it through various angles. It is closely aligned to the detector so that the detector or camera can be rotated in relation to the goniometer. The crystal is mounted on the goniometer head.

The method of mounting depends predominantly upon the stability of the crystal. A stable crystal, of high purity and with well-defined faces which is unlikely to deteriorate quickly in the air is used. It is glued by epoxy resin to a glass fibre which is, in turn, attached to a brass pin by shellac and then placed on the goniometer head. Unstable crystals, such as biological macromolecules, are drawn by capillary action into a thin-walled capillary which is then sealed. They are thus preserved, and drying out is prevented, by immersing in the liquor used in the crystallization process. The capillary is attached to the goniometer head in the same way as that used for a stable crystal.

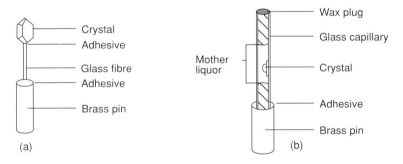

Figure 23 Crystal mountings (a) stable crystal glued onto a sample holder (b) air sensitive crystal in thin walled capillary

The crystal can be rotated about three axes in relation to the goniometer.

The detector is set in relation to the goniometer head such that the angle through which it moves in order to measure the scattered beam is 2θ, twice the Bragg angle, θ.

The diffractometer

The diffractometer (*Figure 24*) records the angle of diffraction, θ, at any lattice plane and the intensity, I, of the diffracted beam. X-rays diffracted by the sample are converted into voltage pulses, the number of pulses being proportional to the intensity of the diffracted beam. The detector can be used either in scanning mode, where it records the 2θ angles of all the reflections, or kept stationary in order to measure the intensity of a particular peak.

RS•C

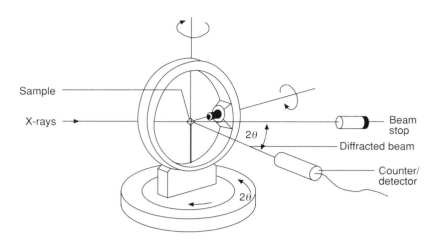

Figure 24 A typical diffractometer

Detecting devices

The detector is usually a proportional counter or a scintillation counter. (The proportional counter is described in the chapter on X-Ray Fluorescence Analysis.)

Scintillation counter **Box 3**

The scintillation counter consists of:

■ an evacuated metal tube;

■ a window admitting the diffracted X-rays;

■ a crystal scintillator;

■ a photocathode; and

■ a series of dynodes.

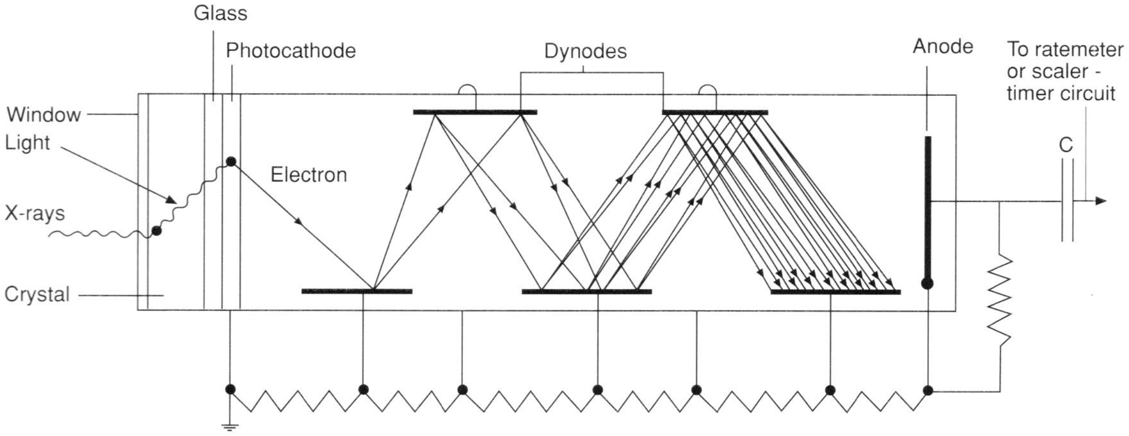

Figure 25 Scintillation counter
(Reproduced with permission from C. Whiston, *X-ray Methods*,
Chichester: John Wiley & Sons on behalf of ACOL, 1987.)

The crystal emits one flash of light each time it interacts with an X-ray photon. A photocathode, acting as a transducer, emits electrons when a light photon impinges on it. The electrons are attracted to the dynodes, each one of which is maintained at a positive potential of at least 100 volts relative to the preceding dynode. As each electron reaches a dynode it ejects more electrons, producing a cascade effect. The anode collects a large number of electrons, and this charge builds up at the capacitor. The charge is detected and is led to an output, either in counts per standard time versus 2θ, or as an average counting rate over a preselected time for a particular value of 2θ.

RS•C

A polaroid film is sometimes used in place of the electronic detector. This records a pattern of spots, the coordinates of which are taken to calculate unit cell size. The intensities are measured with a densitometer. Single crystal analysis can be carried out in a few minutes but, for complex biological molecules such as proteins and nucleic acids, the procedure may take up to two hours.

Finding structures

One factor that has to be taken into account when indexing planes and measuring intensities is that the absence of a reflection does not necessarily mean that no atoms are present or that the atoms are relatively light.

Figure 26 shows the a reflection from the 100 planes of a face-centred lattice of an element.

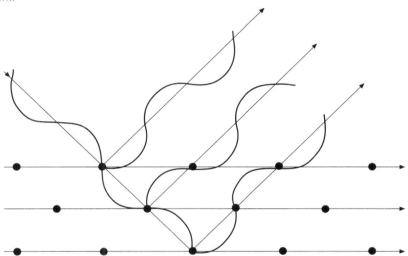

Figure 26 Diagram of interference from 100 planes of an FCC lattice

There is constructive interference between the two 100 planes and the intensity of the reflection is related to the number of electrons in each atom. However, there is another plane (a/2) which is reflecting waves exactly out of phase with the reflections from the two 100 planes. Since there are equal numbers of the two types of plane there is complete destruction and no 100 reflection, although 100 planes are present.

Finding the structure of sodium chloride

This can be illustrated in the determination of the structure of sodium chloride. The empirical data known about sodium chloride are that it is an ionic compound with the empirical formula NaCl and that each unit cell contains a number of Na^+Cl^- pairs or formula units.

The following data are needed:

1. The number of formula units in a unit cell

2. The intensities of each indexed reflection.

Finding the number of formula units

A diffractogram or photograph is obtained by rotating a crystal of sodium chloride about each of its three axes and subjecting it to a monochromatic X-ray beam. Analysis of the photograph or diffractogram suggests it has a cubic structure. The

RS•C

length of the side of the unit cell is equal to 5.631×10^{-9} m.

The volume of the unit cell is $(5.631 \times 10^{-9})^3$ m^3 = 178.5×10^{-27} m^3

To find the number of formula units in this space the following are required:

■ the density of crystalline sodium chloride

■ the molar volume of sodium chloride

■ the volume per formula unit of sodium chloride

■ the number of formula units in the unit cell.

The density of sodium chloride can be found by placing crystals in a mixture of non-polar liquids. A number of mixtures is prepared until one is found in which the crystals neither float nor sink. The density of this liquid mixture is calculated and is equal to the density of crystalline sodium chloride, $(2.16 \times 10^{-9}$ kg m$^{-3})$.

The molar volume $\quad=\quad$ molar mass/density

$\qquad\qquad\qquad=\quad 58.61 \times 10^{-3}$ kg/2.16×10^{-9} kg m^{-3}

$\qquad\qquad\qquad=\quad 27.13 \times 10^{-5}$ m^3

The volume of one NaCl formula unit

$\qquad\qquad\qquad=\quad$ Molar volume/Avogadro number

$\qquad\qquad\qquad=\quad 27.13 \times 10^{-5}$ m^3/6.03×10^{23}

$\qquad\qquad\qquad=\quad 45.0 \times 10^{-27}$ m^3

The number of formula $\quad=\quad$ volume of unit cell/
units per unit cell $\qquad\qquad$ volume of one formula unit

$\qquad\qquad\qquad=\quad 178.5 \times 10^{-27}$ m^3/45.0×10^{-27} m^3

$\qquad\qquad\qquad\sim\quad 4$

Determining the cubic structure of sodium chloride is shown in Box 4.

RS•C

Determining the cubic structure of sodium chloride Box 4

The equation

$$d_{hkl} = a/(h^2 + k^2 + l^2)^{1/2}$$

relates the unit cell parameters to the interplanar spacings and the Miller indices.

The above equation can be combined with the Bragg equation

$$n\lambda = 2d\sin\theta$$

to give

$$\sin^2\theta = \lambda^2/4a^2 (h^2 + k^2 + l^2)$$

which relates the Bragg angle to the Miller indices.

In a primitive cube all possible planes are present, *ie* all reflections are observed. For a face-centred cube, however, not all possible *hkl* values are present because of destructive interference. These absent reflections are called systematic absences. The following reflections and absences are expected. The analogous condition for a face-centred cubic lattic is that h,k,l must be all odd or all even.

hkl	100	110	111	200	210	211	220	221 or 300	310	311
absent	✗	✗			✗	✗		✗	✗	
present			✓	✓			✓			✓

Below are the first few reflections obtained at increasing Bragg angles for sodium chloride. For further information on indexing diffraction data refer to S.E. Dann, *Reactions and Characterization of Solids*, London: Royal Society of Chemistry, 2000.

θ_{hkl} °	$\sin^2\theta_{hkl}$	$(h^2 + k^2 + l^2)$	(hkl)	Strength of reflection
13.41	0.0538	3	111	weak
15.51	0.0715	4	200	strong
22.44	0.1457	8	220	strong

The high intensities of the *200* and *220* planes indicate that they contain all the ions. The *111* planes consist of two sets, one containing only Cl^- ions, the other containing only Na^+ ions. The *111* reflection is produced by alternate equally spaced layers of Cl^- and Na^+ ions.

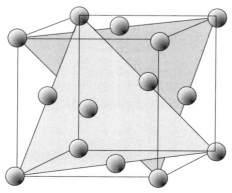

Figure 27 Unit cell for sodium chloride showing *111* plane
(Adapted with permission from L. Smart and E. Moore, *Solid State Chemistry,* London: Chapman & Hall, 1992.)

RS•C

Determining the cubic structure of sodium chloride   **Box 4 continued**

Since the two layers are completely out of phase, destructive interference would be expected and no reflection would appear. However, there is a very weak reflection, because the Cl^- ions are better scatterers of X-rays having 18 electrons compared with 10 electrons for each sodium ion. There is only partial interference. The structure of the unit cell consistent with this data is given in Figure 28.

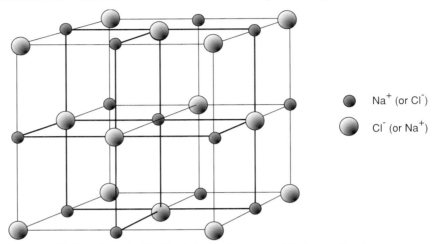

Na^+ (or Cl^-)

Cl^- (or Na^+)

Figure 28 How ions in NaCl relate to planes and to the unit cell

(Adapted with permission from L. Smart and E. Moore, *Solid State Chemistry*, London: Chapman & Hall, 1992.)

Working out covalent structures

Enormous progress has been made with XRC in determining the structures of complex biological molecules. However, the methods, calculations and derivations are very complex. The structure of N-acetylgylcine (N-ethanoylaminoethanoic acid) which was determined in the 1950s provides an example of the procedure. The following stages were involved:

1. The product was synthesised and purified by recrystallisation from water.

2. Batches of crystals were grown from aqueous solution.

3. An X-ray diffraction pattern was collected. The unit cell size was estimated. The spots were indexed and the lengths of the three sides of the unit cell were measured.

4. The density of the substance was measured which enabled the number of molecules per unit cell to be calculated.

5. The reflections from the *100* planes were measured and found to be exceptionally intense, which suggested that all the atoms in the molecule were concentrated in that plane. Cleavage of the crystal parallel to the *100* plane occurs very easily, which supports this idea.

6. The perpendicular distance between the planes was calculated to be 3.24×10^{-10} m. This small distance suggested that the molecules within the layers had a planar arrangement.

7. Trial structures were suggested based on the chemistry.

RS•C

8. The electron density map was interpreted and proved to be straightforward because of the planar arrangement of the monolayers.

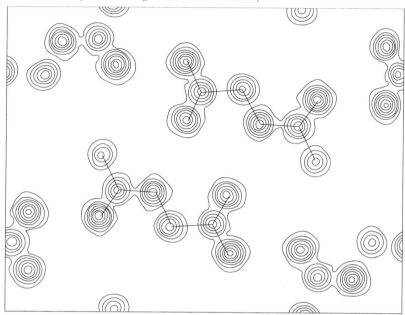

Figure 29 Electron density map of N-acetylglycine (ethanoylaminoethanoic acid)

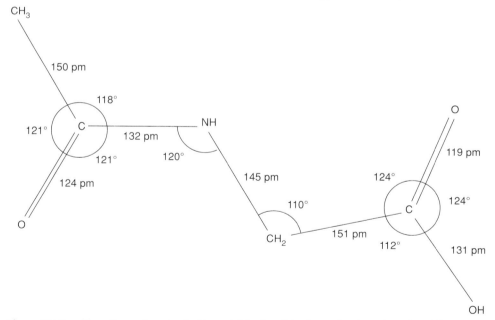

Figure 30 Bond lengths and angles in N-acetylglycine (N-ethanoylaminoethanoic acid)

Computing an electron density map

The data needed to construct a three dimensional electron density map are based on:

■ The amplitudes of the diffracted waves. (These can be calculated because amplitude is proportional to intensity.)

■ The location within the unit cell for which the calculation is done.

RS•C

- The relative phases of the diffracted waves. (This information is lost in the diffraction pattern).

It is this final point that makes the determination of structure such a complex task. If visible light could be used instead of X-rays, a lens could be used to refract the diffracted light waves and form a magnified image of the internal structure of the cell. A lens retains the information on phase changes as light passes through it. No lens has been invented that can focus X-rays, and the best that can be done is to receive the diffracted waves on film or electronically. While this provides information on intensity and position, it conveys no information about phase relationships. Thus, complex calculations have to be done to derive this information.

As has been seen in the case of N-acetylglycine, a theoretical model can be assumed based on what is known of the chemistry of the compound and structures of similar substances. The information is used to compute by extrapolation a diffraction pattern, which can then be compared with that actually obtained. Further calculations and adjustments to structure can be made to make the information fit the data.

Locating the hydrogen atoms   **Box 5**

The diffracting or scattering power of an atom is proportional to the number of electrons, hence X-rays are normally unaffected by hydrogen atoms and pass through relatively undisturbed. There is no data to determine the positions of hydrogen atoms and these have to be inferred. Where it is important to locate the hydrogen atoms in a structure, neutrons are used instead of X-rays. Neutrons are scattered by atomic nuclei and the mass of a nucleus has little effect upon its scattering power.

Neutron diffraction is a complementary technique to X-ray diffraction and is not used on a daily basis due to the need for high energy neutrons generated by a nuclear fission process. This means the cost of neutron diffraction is extremely high and this method is only carried out when data cannot be obtained by any other technique.

RS•C

Powder X-ray diffraction

Powder X-ray diffraction (PXD) measures the Bragg angles and intensities of millions of tiny crystals of a mixture or compound rather than just one pure crystal. The crystals are randomly oriented with their planes in all possible directions so a number of reflections are obtained at the same time. While this method has many advantages, overlap of intensities make it unsuitable for structure determination.

One of the most powerful and widely used applications of powder X-ray diffraction is to 'fingerprint' crystalline compounds. This is particularly useful in identifying geological samples from a mixture and in separating out different crystalline phases in the food industry. Essentially a solid sample is ground down using an agate pestle and mortar, placed into an aluminium holder and smoothed down. If a monochromatic X-ray beam is incident on the sample there is a good chance that some of the crystals will have their planes disposed to fulfil the conditions for Bragg's Law. The reflection can then be detected.

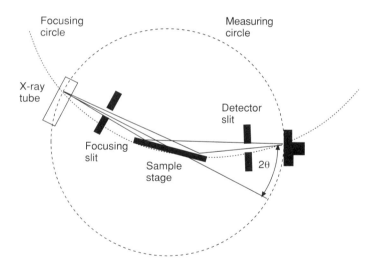

Figure 31 Schematic diagram of a powder X-ray diffractometer
(Reproduced with permission from S.E. Dann, *Reactions and Characterization of Solids*, London: Royal Society of Chemistry, 2000.)

Knowing λ, the wavelength of the monochromatic beam, the *d* values for the interplanar spacings can then be calculated.

RS•C

Carrying out an analysis

The crystals are usually ground down and placed in an aluminium holder. The powder can be pressed down with a glass slide or a spatula. Bulk samples can be examined either by fitting them straight into the sample holder or by filing down.

When taking measurements the detector is driven at a constant speed, and the diffracted X-rays are registered as a series of peaks.

A large number of the powder diffraction data sets collected from inorganic, organometallic and organic compounds have been compiled into a database by the Joint Committee on Powder Diffraction Standards (JCPDS). This database can be used to identify unknown materials by comparing two or three intense lines from the powder diffraction pattern with listings of the most intense lines in book form or more commonly by computer programme.

Texture   **Box 6**

The correct identification of a compound in powder X-ray diffraction depends on all the tiny crystals being randomly orientated, thereby producing a diffractogram accounting for all the possible planes in a crystal. There are cases where crystallites in a sample will have a non-random or preferred orientation and are **textured.** This occurs, for example, when metals are drawn into thin wires.

Texturing influences the properties of a material. Organic fibres, such as muscle, have long molecules wound round the longitudinal axis in the form of a helix. Fibres can be made stronger by inducing preferred orientation. The electrical properties of thin films of titanium depend on the degree of preferred orientation.

Texturing is determined by measuring the variation in intensity of a single Bragg reflection. This is carried out by plotting intensity while the sample is tilted and rotated.

Applications

Since chemical compounds have their own unique interplanar spacings they will have characteristic Bragg angles. This allows for the identification of compounds such as minerals in a geological specimen which cannot be easily characterised by optical microscopy.

Finding the taste for chocolate

Chocolate has to be 'engineered' so that it has an appealing flavour, texture and appearance. Approximately 30% by mass of chocolate is fat and much of this fat consist of cocoa butter which is polymorphic, consisting of six different crystalline forms. Each form has a number, eg Form 1, Form 2 etc and is composed of chains of triglycerides, the main difference between the forms being the distance between the chains. Of these six forms the most stable is Form 6 but its melting temperature of 35 °C is too high for the temperature of the mouth.

Confectioners need to ensure that Form 5, which has a melting point of 32 °C, predominates in their chocolate. At this temperature it will slowly melt in the mouth – an ideal condition for chocolate, known as good mouthfeel. Form 5 crystals are made by tempering. The cocoa butter is cooled just below the melting point of Form 5, warmed up very slowly to the melting point then passed down a cooling tunnel to form the crystals. X-ray diffraction is used to monitor samples of chocolate to check they consist of Form 5 and not Form 6.

RS•C

Industrial dust

Industrial dust is a major factor in respiratory diseases, and one of the major constituents thought to be responsible is crystalline silica. XRD is used to monitor clays and dust and can detect silica in clays down to 0.5% by mass.

Radioactive pipe scale

Hot arid areas such as the Midde East have vulnerable supplies of potable water. The discovery of radioactive material in water pipes in the Jordanian capital city, Amman, resulted in an investigation into the source of contamination. The pipe scale was found to consist of 1.5 ppm uranium. The first question to answer was whether the scale contained any uranium minerals or whether the uranium was adsorbed on to the iron oxides present in the pipes. XRD indicated that no uranium minerals were present. What then could be the source of the uranium?

It is possible that the uranium-rich phosphate deposits in that area may be the source. Uranium minerals are present in veins in the limestone formations underlying the phosphates, and may have been precipitated from fluids seeping down. One idea which may account for the presence of the uranium is that rainwater leaches it from the phosphates, and it is then carried down and concentrated in the underlying structures. At depth, precipitation possibly forms a body of uranium ore. The wells supplying the water are located near these putative uranium ores, and so become saturated with uranium which is adsorbed on to the iron oxides lining the old pipes.

Detecting bentonite

Bentonite is a fine-grained clay which provides a large chemically active surface area. It has an unusual property in that individual crystals are negatively charged but electrical neutrality is maintained by cations located between the layers or crystals. This property, together with its small crystal size, explains its large chemically active surface area and its ability to modify the flow characteristics of liquids.

Some of the more prominent uses of bentonite are listed below:

- Moulding sands for foundry work. Bentonite is mixed with sand and water and formed round a template. When the template is removed to form a mould, hot metal can be poured in and cast.

- Engineering projects. Slurries of bentonite seal porous layers and inhibit water movement in building foundations.

- Pelletising iron ore. Bentonite binds fine-grained iron ores into pellets before they are introduced into the blast furnace.

- Pet litter. The absorbent properties of bentonite make it useful as pet litter.

- Bleaching properties. Bentonite removes colour from edible oils and fats and is used for filtering and cleaning hydrocarbons.

Impurities such as calcite, cristobalite and gypsum detract from the properties of bentonite. X-ray diffraction is used to provide information on the mineralogical composition of bentonite and to determine its purity.

Talc

Talc, $Mg_3Si_4O_{10}(OH)_2$ is a mineral widely used as a cosmetic for absorbing body oils, because of its softness, smoothness, whiteness, lubricant properties, chemical inertness and large surface area. As well as its use in cosmetics, talc imparts whiteness to paints, coatings for paper and acts as a base for ceramic insulators. Commercial grades of talc range from high purity cosmetic grades to low grade material containing other mineral components.

RS•C

Powder X-ray diffraction is used for initial identification of talc and for any minerals that might be associated with it. It can thus be used as an initial assessment of purity.

Dental applications
Most toothpastes contain calcium carbonate as a filler. However, there are different crystalline forms of calcium carbonate which are not chemically distinguishable but do differ in their hardness. It is important to select the correct structure, calcite, as the raw material for incorporation into the toothpaste. This can be done by examining diffractograms of different samples.

Classroom activities
The principle of diffraction can be illustrated using Space spectacles where a spectrum is obtained due to the difference in the wavelengths of the diffracted light. Alternatively simple materials such as cotton will act as a grating. The cotton is stretched and stapled between two pieces of cardboard. If pieces of fabric with different pore sizes are used, it is possible to see the differences between the widths of the banding patterns. An everyday effect of diffraction can be observed by looking at a distant sodium vapour street lamp through an umbrella. Ordinary light shining on the surface of a CD-ROM will diffract to show a coloured spectrum.

RS•C

RS•C

Optical microscopy

Introduction

Optical microscopy has long been regarded as the province of the metallurgists, geologists and biologists. It is a very powerful tool in metallurgy for checking the size and shape of 'grains' in metals such as tungsten in lamps. Geologists use optical microscopes for identifying minerals before rocks are subjected to more detailed analysis using X-ray diffraction or electron microscopy. Optical microscopy is used in biology, in studying living tissues and small plants and animals. However, this technique is now being used in diverse analytic and diagnostic ways in chemistry such as:

■ diagnosing faults in paper during manufacturing;

■ checking the uniform lacquering of tin plate;

■ checking the thickness of coatings on tablets in the pharmaceutical industry;

■ examining small surface sections on steel bars for evidence of corrosion; and

■ analysing the surfaces of fabricated plastics which fail to adhere after welding.

In these examples a section of the object is viewed either through the eyepiece of a microscope, or with the help of a camera attached to the microscope. Most objects can be studied over a long period of time because they do not change much. Recent advances in optical microscopy allow dynamic processes to be studied in real time.

The principle of a microscope

The purpose of a microscope is to help the eye to see objects in greater detail and to provide a magnified image of the object. A single lens used as a magnifying glass is the simplest example of a microscope (Figure 1). It works by increasing the apparent angle that the object makes with the retina of the eye.

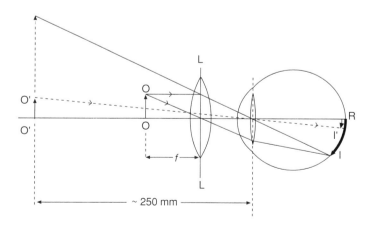

Figure 1 A ray diagram illustrating the principle of the magnifying lens
Two points on an object O'–O' produce a small image on the retina of the eye (I'–R) and the points are not resolved – ie they cannot be seen as two distinct points. If the object (O–O) is then moved to the focal point of the convex magnifying lens (L) the visual angle is increased producing an erect, virtual and magnified image at infinity. The retinal image (I–R) is now resolvable.

RS•C

The more common arrangement is the compound microscope (Figure 2), which has at least two lens systems – the objective and the eyepiece.

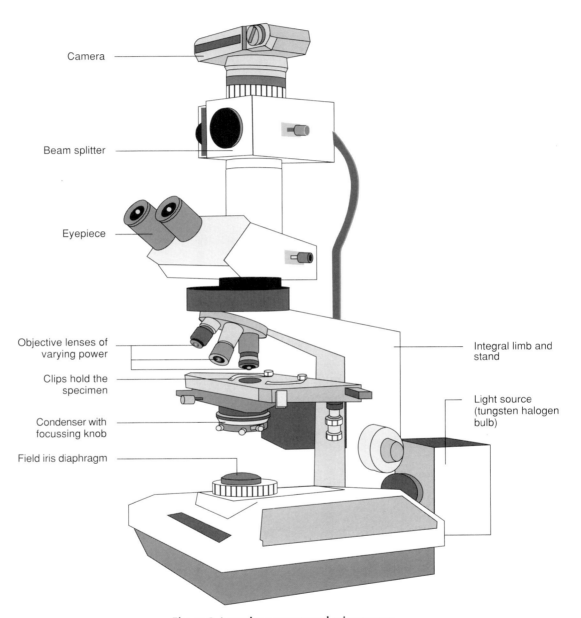

Camera

Beam splitter

Eyepiece

Objective lenses of varying power

Clips hold the specimen

Condenser with focussing knob

Field iris diaphragm

Integral limb and stand

Light source (tungsten halogen bulb)

Figure 2 A modern compound microscope

The components are arranged around an integral limb and stand. The light source is a tungsten halogen bulb. The light passes through a set of collecting lenses before reaching the field iris diaphragm. This can be controlled to limit the light that reaches the rest of the system. The light reaches the condenser whose height can be varied using a focusing knob. Clips hold the specimen on the platform. The resultant light from the specimen-light interactions is collected by the objective lens carrying the required magnification. The image reaches the eyepiece. The microscope is adapted for taking a photo of the image (photomicrograph) with a camera and a beam splitter to allow simultaneous viewing through the eyepiece and camera.

RS•C

The objective lens is placed nearest to the object from which it collects reflected and transmitted light, and forms the primary image. The eyepiece magnifies the primary image formed by the objective lens and produces a virtual image for viewing. The condenser lens is a lens used underneath the stage of the microscope to concentrate light on the specimen (Figure 2). (The specimen refers to the object being viewed under the microscope.)

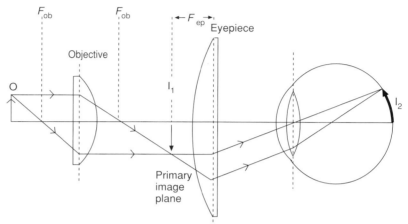

Figure 3 A ray diagram to illustrate the relationship of objective lens and eyepiece in a simple compound microscope. When the object O is placed just outside the focal point of the objective lens a real, magnified and inverted image, I_1, is formed at the primary image plane. When the eyepiece is moved so that the primary image plane is at its focal point, the final, magnified image (I_2) is formed.

As well as magnification, a microscope should also improve contrast. This means that individual features of an object can be distinguished because the image is differentiated into light and dark areas, different colours or some other form of contrast.

Contrast systems in optical microscopy

Most optical microscopy does not use added filters in the light path. The contrast observed in the image results entirely from the interactions of the specimen with light. This is known as brightfield microscopy. It is used to focus and align the microscope before any other contrast system is used.

Brightfield microscopy gives an idea of the physical organisation of a system but gives no chemical or detailed structural information.

RS•C

The photograph shows a brightfield image (Figure 4) of a fabric softening solution.

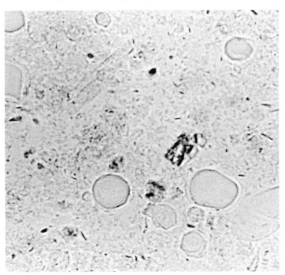

Figure 4 A brightfield image of a fabric softening solution

An alternative method of image formation is darkfield microscopy. In this type of microscopy the object is illuminated from an angle such that no light from the source reaches the objective lens directly. This is achieved by placing a 'patch-stop' (Figure 5) in the light path between the light source and the condenser (lens) under the microscope stage.

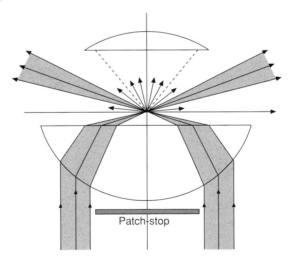

Patch-stop

Figure 5 The principle of darkfield illumination. The central patch-stop in the condenser provides a hollow cone of light peripheral to the central stop. The light scattered or diffracted by the specimen forms an image in reverse contrast.

The images observed in darkfield microscopy have the reverse contrast to brightfield microscopy. Large objects appear as light outlines, but small objects and fine details appear as bright objects against a dark background. Consequently darkfield microscopy can be used to show detail that is lost in brightfield microscopy, such as cell boundaries or the outline of small crystalline structures. Darkfield microscopy also gives fine detail, which is not available in brightfield microscopy (Figure 6).

RS•C

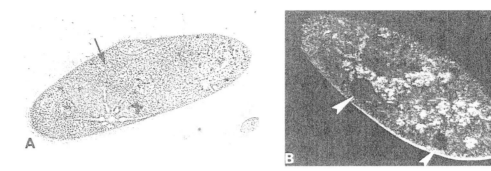

(a) brightfield   (b) darkfield
Figure 6 The difference between brightfield (a) and darkfield (b) from a photomicrograph of the single celled animal *Paramecium caudatum*

Another form of microscopy, polarised light microscopy (Figure 7), offers a refinement giving rather different images in some cases. The images are from optically non-uniform specimens, where the non-uniformity shows up as coloured streaks.

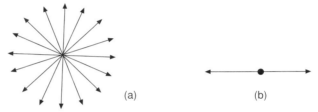

(a)   (b)

Figure 7 The action of a polariser. Consider the light beam emerging towards you through the plane of the paper, its waves vibrating in all possible directions (a). After passing through a polariser, the light is plane polarised with light waves oriented predominantly in a single plane (b).

To achieve these images, the specimen is placed in the light path between two polarising filters. When light passes through a polarising filter – *eg* polaroid – the light becomes plane-polarised (consisting of light in one plane only). The molecules of the polarising filter are oriented in such a way that only light predominantly in a single plane can pass through (Figure 8).

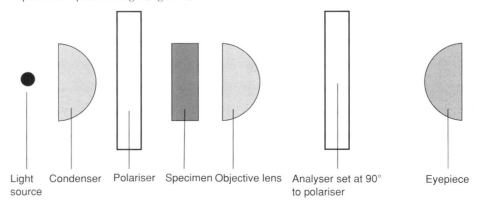

Light   Condenser   Polariser   Specimen Objective lens   Analyser set at 90°   Eyepiece
source   to polariser

Figure 8 The arrangement for polarised light microscopy

Most brightfield microscopes can be adapted to give polarised light images. One polarising filter – the polariser – is placed on the lens over the base of the sub-stage condenser so that light from a source is plane-polarised. The second filter – the analyser – is put between the objective lens and the eyepiece and rotated until the background is as dark as it will go. The two polarising filters are now 'crossed', with the planes of light that they transmit at 90° to each other.

RS•C

When the sample is placed in position in the microscope, any image that appears does so because the sample has rotated the plane of the polarised light. If the sample is anisotropic – *ie* optically non-uniform, the light coming from the polariser is changed. Crystals (Figure 9), plastics and other polymers split the polarised light into components that are out of phase with one another. This phenomenon is known as birefringence.

Figure 9 The comparison of photos of vitamin B crystals – brightfield on the left and polarised light microscopy on the right

The analyser brings the out-of-phase light rays back into one plane but the two rays have interfered with one another. This means that some components of the white light spectrum have been diminished because the resulting interference has reduced the amplitude and hence the intensity of the waves entering the eyepiece (Figure10). This results in colour contrast which can be seen in the vitamin B crystal (Figure 11).

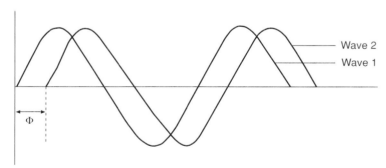

Figure 10 A diagram of two light waves with the same amplitude but different phase
Wave 2 is advanced by a distance Φ relative to wave 1. This cannot be detected by the human eye but is changed optically into an amplitude difference, which is detectable in the phase contrast microscope.

RS•C

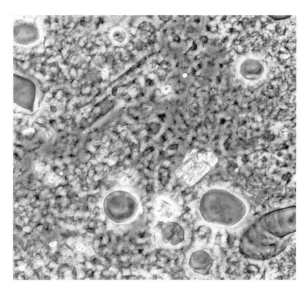

Figure 11 A photograph of the same fabric softening solution as in Figure 4 taken under phase contrast conditions

The principle of phase contrast microscopy involves light interference to improve contrast. When light rays pass through a transparent specimen they are not absorbed and their intensity is not diminished. However, the chemical components within the specimen refract – *ie* slow down – the rays differentially and interference occurs due to the phase changes produced.

The phase changes are very small and the effect is insignificant in ordinary compound microscopes. Phase contrast microscopy increases the phase differences which, in turn, leads to greater contrast. It is particularly useful for specimens with little optical contrast (Figure 11).

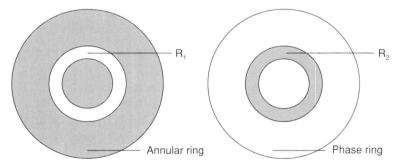

Figure 12 A diagram of an annular ring and a phase plate in a phase contrast microscope

Phase contrast microscopy has a similar arrangement to that for brightfield microscopy but there are modifications in the objective lens and the condenser. The objective lens has a phase ring in the phase plate. This consists of a transparent substance, such as magnesium fluoride, which further slows down the light. A ring-shaped diaphragm – the annular ring (*Figure 12*) – is placed in the condenser. The images then have haloes around them. Bright haloes enshroud dark images against a lighter background and dark haloes surround bright images against a darker background.

The refracted light travels to the objective lens and can only pass through the transparent part of the phase plate. Since there is a great reduction in light passing

RS•C

from the specimen it is important that the transparent area of the annular ring is completely covered by the opaque ring in the phase plate.

Phase contrast microscopy can be used to estimate the film thickness of polymers. Interference patterns of recombined rays correspond with variations in the thickness of the polymer film. The photograph of a fabric conditioner shows how different areas can be identified, compared with brightfield and polarised microscopy.

Differential interference contrast microscopy (DICM)

This type of microscopy produces images with a relief appearance, rather like looking down on a mountain range from a distance. It is an advance on phase contrast systems. The light is polarised first then it passes through a prism in the condenser, which is used to split the beam in two. One beam passes through a detail of the specimen and the other one passes to one side. Another prism recombines the beams after they have passed through the objective lens.

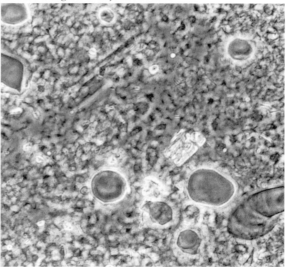

Figure 13 A photograph of a fabric softening solution taken under differential interference contrast microscopy (note the 'relief' effect)

The phase change results as a difference in the light interaction with the specimen. Collecting the diffracted light through the beam recombiner reveals details that are not always available using other microscopical systems. Again compare the photograph of the fabric softening solution with that obtained from brightfield microscopy.

Video enhanced microscopy (VEM)

This technique involves a video camera used in conjunction with a light microscope In addition to the fact that viewing a TV monitor is easier than looking down the eyepiece of a microscope, the major advantage is that VEM overcomes the inability of the human eye to detect small contrast differences in an image, particularly where there is low contrast and high brightness. Video cameras can offset a bright background enabling small changes of contrast to be seen. There are several advantages to VEM.

■ Faint structures and subtle differences in microstructure can be seen.

■ Viewing and analysis is easier because the image is enhanced on a TV screen.

■ The image can be computer processed through an image analyser so that certain structures can be highlighted and quantified.

RS•C

- Very low contrast changes can be detected.

- Complex 3-dimensional surfaces can be probed.

- Observations can be made in situ.

- Dynamic systems – *eg* the effects of bleaching agents on fabric – can be investigated.

The kinds of properties that can be investigated using VEM are the distribution and size of particles, mechanical properties, dispersion relationships of components, temperature and storage effects, and their relationships between sensory characteristics such as the 'feel' of a cream, 'odour' and so on.

Image analysis

Image analysis involves using a TV camera linked to an optical microscope. The image is digitised and stored. The possibilities of investigating the image are enhanced by the fact that it consists of many pixels, each of which can be represented at 256 grey levels. A pixel is a small discrete element that, with other pixels, constitutes a whole picture. In terms of a digitised picture, a pixel is one of the dots or resolution elements making up the picture. In this computerised form the images can be treated, enlarged, distorted and many sub-structures can be analysed and quantified. The image can be enlarged to 1024 x 1024 pixels; the greater the enlargement the greater the resolution.

The controlling software can manipulate the image and highlight certain features.

Fluorescence microscopy

Fluorescence microscopy is a technique that depends on light of one wavelength exciting a material. The material then emits light of a longer wavelength. The incident light is sometimes in the ultraviolet region, and therefore invisible to the naked eye. For this reason fluorescent objects may appear to be self-luminous.

Some specimens are naturally fluorescent – *eg* teeth. Others can be made to fluoresce by introducing fluorescent tags, such as acridine orange. The fluorescent components appear bright against a dark background, which gives sharp contrast that enables the location of chemical species and the identification of sub-structures. As an example, the distribution of the organic and aqueous phases within lipstick can be examined in combination with differential interference contrast microscopy (DICM).

The optics of a compound microscope have to be modified for fluorescence imaging – *ie* filters need to be added. Excitation filters allow only those wavelengths to pass through and bring about fluorescence. Barrier filters prevent light getting to the image which cause fluorescence but allow light produced by the fluorescence of the specimen to pass through. The lenses must also be non-fluorescent. Four different light sources can be used – tungsten halogen, high pressure mercury, high pressure xenon, and a range of lasers.

Confocal microscopy

One problem with the optical microscopes discussed so far is the background 'noise' from light scattered by those parts of the specimen that are not in focus. For example, when focusing on a fibre or small crystal, light from the surface forms a clear image. However, when focussing deeper the object loses contrast due to light scattered from the surface and other structures. This is a problem when looking at distributions of structures in relatively bulky objects, such as living tissue. One way of probing depth in living tissue is to kill the material then section it.

RS•C

Confocal laser scanning microscopes (CLSM) solve the problem of killing or immobilising specimens. It is a non-destructive way of producing very clear images with fine detail of three-dimensional structures. It is often used with fluorescent labelling. When linked with image analysis a series of images is produced that can be stored and used for three-dimensional reconstructions. The distribution of a chemical throughout a material can be traced so that the route of certain molecular species can be followed.

In CLSM all light is eliminated other than that arising from the specific focal plane – the plane of the part of the specimen that is in focus, providing very high contrast (*Figure 14*).

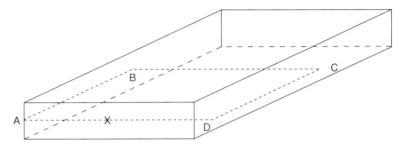

Figure 14 ABCD represents the focal plane of a point X within an object of depth

A point source from a laser, transmitted through a pinhole, illuminates a very small volume of the specimen. The scattered light or fluorescing light is then refocussed on to the focal plane in the sample by the objective lens (Figure 15).

RS•C

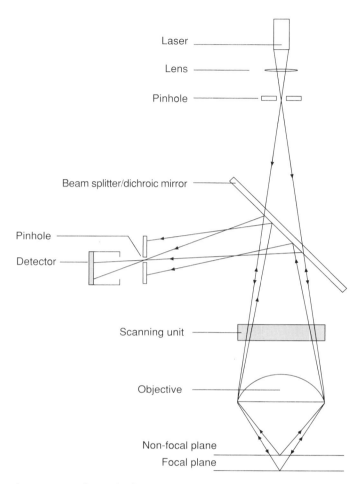

Figure 15 A schematic diagram of the principle of a confocal microscope

There is no blurring from unfocussed points because the focal plane is the area where the image is sharpest. The light from the focused point is reflected from a mirror and passes through a second pinhole towards the detector. The light from unfocussed points does not pass through the detector. The objective lens is used to refocus the point source of laser light on to the focal plane of the volume of the sample being investigated.

A series of sections is taken at different depths within the sample and a three-dimensional image is reconstructed from thick samples. In modern instruments the signal from the detector does not form an image, but is recorded, a point at a time, by a photomultiplier tube. The image is scanned and viewed through a monitor.

Confocal laser scanning microscopy is particularly effective because the detector and source pinholes are conjugated – *ie* only reflected light from the laser enters the detector, so that any unfocussed light is not detected.

Applications of confocal microscopy

Dyeing fibres. How far does a dye penetrate a fibre? Does the dye tend to stay on the surface? Does it penetrate more deeply and, if so, how quickly? Can it be removed quickly? What effect does its removal have on the material?

A fluorescent image of a dyed nylon fibre taken through a compound microscope arrangement shows many regions that contain tiny crystals in a disordered

RS•C

arrangement – shown as bright spots. The light scattering makes it very difficult to obtain information about the distribution of the dye. CLSM shows that initially the dye is adsorbed on to the surface of the fibre, with little diffusion into the bulk. After careful heat treatment the dye slowly diffuses inside. The distribution of the dye can be seen in the CLSM sections despite the fact that a thick region of fibre has been viewed. The microcrystalline regions show up as darker patches. The dye penetrates the amorphous regions of the fibre much more efficiently than the microcrystalline regions.

Suggestions for study

Many household materials provide interesting images when viewed through a microscope using brightfield illumination at reasonably low magnifications (objective 20–40x). The specimens should be placed on a microscope slide.

Crystallisation from solutions. Make up a concentrated sugar solution, place a few drops on a microscope slide and allow to crystallise. View the crystals under a microscope. If a polarising microscope is available the crystals are seen in different colours. The same can be done with household salt and compared with tears – try an onion. Tears crystallise in branched forms, rather like a bunch of flowers. Experiments can be done to see how the rate of cooling affects the shape of crystals that are formed. Try different household solutions and watch out for any interesting crystal formations.

Fibres. Fibrous materials such as hair, paper, animal fibres, cotton and silk should be placed on a microscope slide and covered with a second slide to flatten the fibres. The difference in appearance between a dirty hair and a freshly shampooed hair can be investigated.

Hand creams. Vaseline skin cream is a good sample to study.

Low fat spreads. For example Flora margarine can be investigated. In polarised light the individual fat crystals can be seen.

Tea leaves and tea bags. The relative diameters of the tea leaves and holes can be measured.

Toothpastes. The crystals in the paste should be visible.

Further study

Dynamic studies can be done on samples on the microscope slide. For example, the action of a liquid on a specimen can be investigated by placing a cover slip over the specimen and introducing the liquid at one side of the cover slip. Capillary action then carries the liquid to the specimen. The action of a detergent solution on a fibre covered with – eg olive oil can be observed in this way. Similarly, the way a washing powder granule dissolves in water can be seen.

RS•C

Sampling and sample preparation

Introduction

Making measurements on a sample is only one step in an analysis (Figure 1).

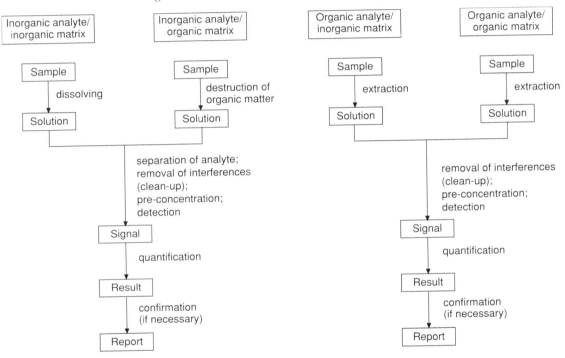

Figure 1 Some of the stages during an analysis

Before taking measurements the sample has to be taken from a larger batch of material, homogenised, then possibly stored and transported before preparation.

Analysing milk illustrates some of these procedures. Stored milk has a layer of cream on the surface so that the fat content is unevenly distributed. The fatty lumps are broken up and mixed with a clean wire whisk. Milk is rarely consumed straight after sampling and, as a perishable, changes are likely to take place in its composition even on refrigeration. These changes include the fermentation of lactose sugars and the degradation of the milk proteins to ammonia. When sampling and analysing milk, corrections must be made for these changes.

Preparation of the sample depends on factors such as the analytical instrument being used, the data that are needed, the cost of the analysis and what the results are being used for.

RS•C

Questions an analyst needs to ask   **Box 1**

Where has the sample come from?   How was the sample taken?

What is the purpose of the analysis?   Which analytes are to be determined?

Is the analysis to be quantitative or qualitative?   What is the likely concentration?

What detection limits are required?   What level of accuracy is required?

What other components are present in the matrix? What are the likely causes of interference?

What form of sample preparation is necessary?   What are the limitations on cost, time and expertise?

Which technique should be used?   How should the data be presented?

How will the results be used?

Sampling

The sample being analysed should provide information about the batch as a whole. For example, suppose 1 g of a sub-sample is to be analysed from a larger 1 kg sample, then the amount of analyte contained in the sub-sample should be 0.1% of the amount of analyte in the larger sample. If the sample is representative then the parent material must be homogenous. Most materials are heterogeneous so the first step in sampling is homogenisation. For solids this includes chopping, milling and grinding. Liquids are mixed with clean implements like whisks or spoons or an electric blender.

Knowledge of both the purpose of the analysis and of the history and nature of the sample is important because straightforward homogenisation can conceal the problem to be solved. The case studies below indicate some of the complex sampling problems facing analysts.

Aflatoxins

Aflatoxins are highly toxic compounds produced by moulds growing on nuts in hot humid atmospheres. The moulds may only appear after the nuts have been harvested and shipped off, but they are usually localised on damaged kernels. When samples of nuts are taken for analysis from a consignment, infected nuts can be missed and the batch passed as fit for consumption.

Monitoring raw groundnut kernels in the US requires a minimum sample of 66 kg from each batch by taking every fourth sack which is subdivided into three equal portions of 22 kg. These sub-samples are then divided into 20 equal portions and ground to produce a 1.1 kg portion. After analysing these portions the batch is accepted, rejected or analysed further. In the UK an analytical result below $10 \ \mu g \ kg^{-1}$ of aflatoxin is considered to be fit for consumption.

The problem facing public food analysts is immense, given the huge amount of nuts and nut products imported into many countries, because there are not enough staff to carry out systematic inspection on all shipments. As a result goods from high-risk regions are usually targeted.

Tea

Atomic absorption spectrometry (AAS) is a well-established technique for determining the concentrations of metals, such as lead, in foods and drinks. Samples taken from homogenised samples of packeted tea show a surprising variation in lead content (Figure 2).

RS•C

Sample	Pb found /mg Kg^{-1}	
A	6.2	1.1
B	1.2	6.6
C	3.9	3.9
D	25.0	6.5
E	3.9	9.5
F	4.2	2.4
G	<0.2	0.6
H	0.6	0.4
I	<0.2	1.4
J	0.6	0.4

Figure 2 Data on different results of lead in tea

The explanation for the discrepancy lies in the difficulty of obtaining truly representative samples of packet tea. Each sample tested contains a mixture of dust and leaves, where the ratio by mass of dust to leaves (Figure 3) varies considerably.

Sample	% Leaves	% Dust	Lead Leaves	Lead Dust	Iron Leaves	Iron Dust	Zinc Leaves	Zinc Dust	Copper Leaves	Copper Dust
A	44	18	0.2	17.2	150	1300	35	105	17	32
B	55	18	5.9	14.6	350	1500	40	140	20	33
C	33	19	0.4	8.5	150	1350	35	100	19	33
D	33	21	0.2	11.7	300	1400	35	125	20	33
E	34	17	8.6	10.0	200	1200	140	120	20	.31
F	33	19	0.2	6.0	150	1400	30	100	20	34
H	78	6	0.2	3.7	200	950	35	70	26	33

Figure 3 Distributions of metals between leaf and dust fractions of tea

Fractionating the samples by sieving shows the concentrations of lead in the dust to be far higher than in the leaves.

This raises the problem of deciding what should be sampled. Legal limits on lead content relate to infusions, not to the tea leaves or dust. Only about 10% by mass of the lead in the dry matter is infused, but a soluble component in tea forms a complex with lead which makes it difficult to extract the lead and prepare it in a form suitable for analysis.

River pollutants
Polynuclear aromatic hydrocarbons (PAHs) are a pollutant in river water in concentrations of ng dm^{-3}. When river water is filtered the concentrations of PAHs detected are far lower than in unfiltered river water because PAHs concentrate in mud particles in the river sediment. Knowledge of the distribution of an analyte can determine how the sampling and the subsequent analysis is carried out.

Meat
Tetraethyllead(IV) concentrates in the strings of fat in meat rather than in the leaner flesh. Its effect on health depends on how the meat is treated, cooking procedures and the eating habits of consumers. Homogenisation of samples of meat may therefore give results that are deceptive for people who tend to eat the fatty parts of the meat.

RS•C

Types of sample

Poor sampling increases the risk that sub-standard batches of material are accepted and perfectly good batches are rejected. International protocols for good sampling practice have been instituted for many food products and pollutants.

The ideal situation for sampling is a completely homogeneous system where any part of the whole is suitable for analysis, but this is rarely attainable. Hence, the main question is how to obtain a representative sample from a heterogeneous material. This is a particular problem for solids, which are often heterogeneous and which, for soils, clays and rocky materials, may need drastic treatment to be made more homogeneous (Figure 4).

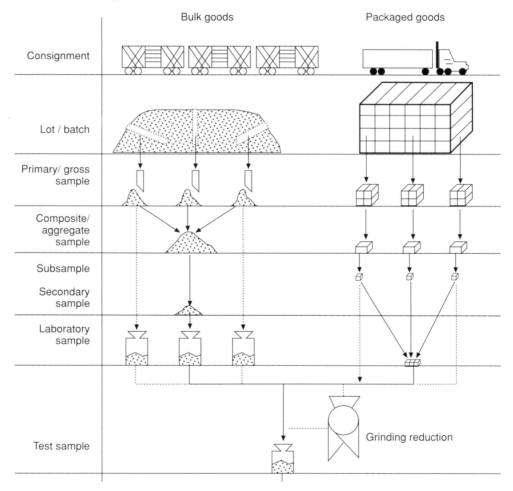

Figure 4 Diagram to show sub-sampling in solids, coning and quartering

However, solids are relatively static. For flowing river water it is impossible to decide whether a sample is representative because the system is changing all the time.

One approach is random sampling which provides a basis for the statistical interpretation of measurement data. The principle of random sampling is that any portion of the batch has an equal chance of selection. One example is to select the first sample at random and then take subsequent samples at regular intervals – *eg* every tenth can of baked beans in a supermarket warehouse. In a systematic scheme samples are selected on the basis of a fixed time pattern.

RS•C

Taking the sample

Methods of taking a representative sample depend on the type of material that has to be extracted.

Type of material	Problem	Solution
Particulates on a conveyor belt.	The size of the particles vary in their position on the belt.	A rotating scoop at the end of belt to collect falling particles (Figure 5).
Particulate matter at rest.	The whole batch is not accessible to the sampler.	A long hollow tube with a handle is inserted into the sample and withdrawn. (Figure 6).
Compact solids – eg rocks, metal ingots and concrete.	Samples cannot be obtained without mechanical crushing or drilling.	A drill is used and all shavings and powder ejected from holes collected.
Liquids in flowing systems – eg rivers and industrial effluent.	The chemical composition at any one time varies because of factors such as temperature, the distance from the source, depth and flow rate.	Weighted glass bottles can be left at different depths. (Figure 7). Frequent and duplicate sampling is required to reduce uncertainty.
Liquids flowing in pipes – eg oil.	For slow-flowing liquids the flow-rate at the centre of the pipe is faster than that at the edges.	Create turbulence before a sample is taken (Figure 8).
Liquids in closed containers.	In a mixture containing components of differing densities, stratification occurs. It may not be possible to homogenise the mixture.	Samples are taken at various depths using a plunger device (Figure 9).
Liquids in open bodies – eg a lake.	There are no particular problems with sampling, but difficulties may occur if sampling has to be done on a permanent basis.	Use a permanent sampling device (Figure 10).
Gases in pipes – eg gas piped for domestic purposes or as industrial feedstock or products. Gases below atmospheric pressure need to be pumped.	The composition and mass of the gas depends on temperature and pressure.	Sampling probes can be built into pipework – the temperature of the probes is controlled.
Atmospheric gases.	Changing composition. Interference of natural phenomena such as wind and rain.	Simultaneous sampling at a number of locations. Continuous sampling over a long period – eg 24h.

Table 1 Taking samples

RS•C

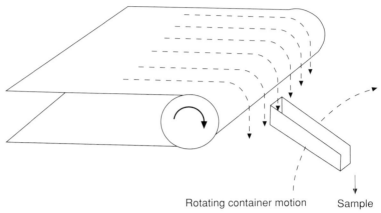

Rotating container motion   Sample

Figure 5 Sampling from the end of a conveyor belt

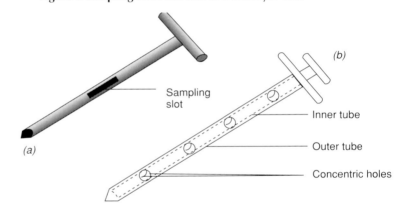

Sampling
slot

Inner tube

Outer tube

Concentric holes

(a)

(b)

Figure 6 *(a)* **Bayonet or split tube thief** *(b)* **concentric-tube sampler**

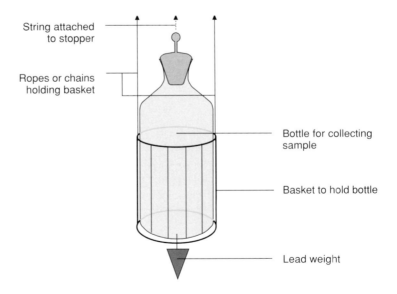

String attached
to stopper

Ropes or chains
holding basket

Bottle for collecting
sample

Basket to hold bottle

Lead weight

Figure 7 A sampling device for liquids in an open system

RS•C

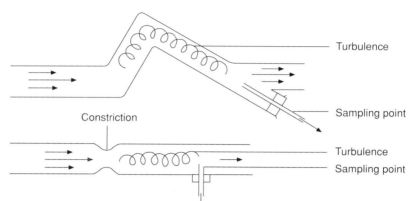

Figure 8 Sampling devices for liquids flowing within closed systems

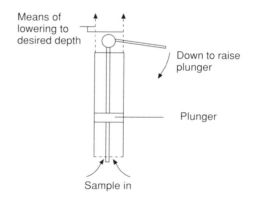

Figure 9 A sampling device for a stratified or inhomogeneous liquid

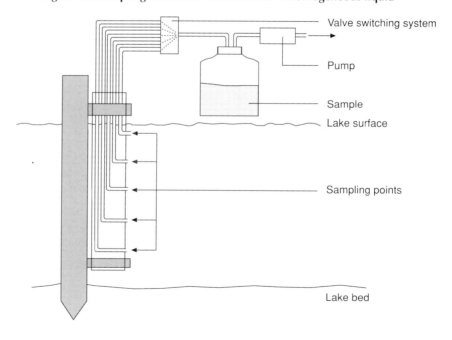

Figure 10 A permanent sampling device for open body of liquid

RS•C

Storage and transport

The best procedure after taking a sample is to analyse it directly. For example, on-line analyses for quality control in the manufacture of cement involve constantly monitoring the composition of the cement slurry by X-ray fluorescence (XRF). However, in most cases the sample has to be stored and transported to a laboratory, sometimes across the world. This can involve changes in climatic conditions with the result that the sample's properties change and it is no longer representative of the batch. The types of changes in chemical composition that can take place are:

- thermal decomposition;

- reactions between sample components;

- changes due to light; and

- small quantities of metal ions acting as a catalyst.

The samples should be stored or preserved in appropriate conditions to maintain their original state as long as possible.

Storage conditions	Samples
Deep freeze	Chemically unstable analytes Bio-organic analytes with high enzymic activity
Refrigerator	Fresh fruit and vegetables Soils Aqueous samples
Room temperature (dark)	Dry powders Minerals
Desiccator	Hygroscopic (water sensitive) samples Pharmaceutical powders and tablets

Table 2　Examples of storage conditions

Containers

Vessels containing samples must be clean and free from contaminants. The two main types of container used are glass and polythene. The problem with liquids and moist solids is leaching substances from the walls of the container. When analysing for trace amounts of boron in fertiliser, storage in borosilicate containers can affect the result because more boron may leach from the walls than the amounts of boron actually present in the sample. As well as metals leaching into the sample from glass vessels, other metals can be adsorbed from the sample on to the glass walls.

Glass vessels should be thoroughly cleaned. This involves acid-washing containers to ensure the complete removal of trace metals. Phosphate detergents may be used, but phosphorus analyses should not be done in these cleaned vessels.

Liquid hydrocarbons should not be stored in polythene bottles. As a general rule, inorganic analytes should be stored in polythene vessels and organics in glass bottles. The exception is mercury, which reacts with organic materials. Where only qualitative identification of metals is needed, and the metals are likely to be in greater than trace amounts, then glass containers are acceptable.

RS•C

Sample preparation

Once a sample has been extracted it is likely to need some preparation before being analysed. This preparation depends on a number of factors including:

- the nature of the analytical technique – *eg* atomic absorption spectrometry, X-ray fluorescence, gel electrophoresis;

- the analyte matrix (is it an aqueous solution, detergent powder, a piece of meat or a lump of rock?);

- the form of the analyte (is it a solid, a liquid or a gas?);

- whether the sample is organic or inorganic?

- the purpose of the technique (qualitative/quantitative, and the level of accuracy); and

- the speciation; the form of the analyte – *eg* arsenic can form part of an inorganic or organic compound.

Complicated preparations, involving many stages and different pieces of equipment, increase the risk of contaminating or losing the analyte. There are some techniques where there is little or no sample preparation which generally produce qualitative information only. Examples are the identification of metals in steel ingots using X-ray fluorescence (XRF) and minerals in rock mixtures using X-ray diffraction (XRD). Bulk solids must be polished to produce an optically flat surface when using XRF. In optical microscopy, a metallurgical or geological sample may have to be microtomed – *ie* cut into a thin section – to produce a section for viewing.

The type of preparation depends on the analyte and the matrix. For convenience these fall into three groups:

- separating an inorganic analyte from an inorganic matrix – *eg* metals in rocks;

- separating an inorganic analyte from an organic matrix – *eg* the environmental analysis of heavy metal concentrations in plant material, and metals from an engine casing in fuel; and

- separating an organic analyte from an organic matrix – *eg* pesticide residues in animal tissue.

Occasionally, an organic analyte may have to be extracted from an inorganic matrix – *eg* oils from a rock sample, but these extractions are rare.

Separating inorganic analytes from inorganic matrices

Common inorganic matrices are metals and alloys, metal ores, silicates, and dry ash residues of plastics, paper and paints. The analyte, usually a metal, is dissolved in water or acid and filtered. Some form of preconcentration is necessary for trace elements to ensure that the concentration of the analyte is above the instrument's lower detection limits. Ideally, complete dissolution of the analytes is needed to form a clear aqueous solution, free of contaminants and within a suitable concentration range. Occasionally, diluting the extracted analyte ensures that the response is within the instrument's linear range.

RS•C

The main acids used are:

■ Nitric acid (HNO$_3$) and hydrochloric acid (HCl), or aqua regia (3:1 v/v hydrochloric acid : nitric acid). These dissolve most metal analytes, though hydrochloric acid is not used to dissolve metals that form insoluble or volatile chlorides. Concentrated nitric acid is an oxidising acid and can be used to dissolve metals and alloys such as stainless steel which are impervious to attack by other acids. Aqua regia dissolves noble metals, including gold and platinum;

■ Hydrofluoric acid (HF). This is used for very intractable matrices such as silicate rocks. Hydrofluoric acid is very corrosive and attacks glass, so dissolutions are done in vessels that are resistant to attack, such as those covered with polytetrafluoroethene (PTFE).

Fusion

This technique is used for rocks, cements, ceramics, slags and other aluminosilicates. The sample is ground and mixed thoroughly with the flux – an acidic or basic electrolyte. (Molten electrolytes are extremely powerful solvents). Successful fusion depends on the correct ratio of flux: sample. The mixture is heated in a platinum crucible until the flux melts – the sample dissolves to give a clear melt. The solution is then cooled, when it can be used for analysis as a homogeneous solid solution in X-ray fluorescence (XRF), or dissolved in acid.

Some common fluxes are:

■ sodium carbonate (Na$_2$CO$_3$), a basic flux for silicates;

■ sodium peroxide (Na$_2$O$_2$) which acts as a basic oxidising flux for sulfides and alloys of iron, nickel, chromium and tungsten; and

■ disodium tetraborate decahydrate (borax, Na$_2$B$_4$O$_7$.10H$_2$O). This is used for aluminium and zirconium ores, rare earth minerals and slags. It is usually mixed with sodium carbonate and can be used at temperatures as high as 1200 °C.

Preconcentration

Once metals have been extracted from a matrix they may have to be preconcentrated. This normally involves evaporating the solvent and redissolution in a more appropriate solvent – eg from HF to a dilute mineral acid. For some volatile metals such as arsenic, antimony and selenium, hydride generation for atomic absorption spectroscopy is a means of removing the metal from its matrix and presenting the analyte in almost pure form for analysis.

Chelation, followed by solvent extraction, is a way of preconcentrating transition metals for inductively coupled plasma – atomic emission spectroscopy (ICP–AES). The sample is complexed with ammonium pyrolidinedithiocarbamate. In this form the metals can be separated by shaking with an organic solvent – eg 1,1,2-trichloro-1,2,2-trifluoroethane. The separated transition metal complexes are then hydrolysed by acid and back-extracted into the aqueous phase for analysis.

Ion exchange columns are one way of removing unwanted metal ions – eg by passing the solution through a column packed with a resin containing iminodiethanoic acid (HN(CH$_2$CO$_2$H)$_2$) which can form chelates with alkali metal ions.

Leaching

This involves taking one or more components of the sample into solution. For example, metals in the soil may be leached directly, or after ashing.

RS•C

Case study

Flame atomic absorption spectrometry (FAAS) was done to quantitatively determine the cobalt uptake in plants growing in a small plot of land. It was assumed that plants take up only soluble forms of the metal, so leaching conditions were made as close as possible to those of the plants' ability to extract metal ions.

A sample of soil was weighed and dried to constant mass then shaken in a solution of ammonium ethanoate, which acts as a buffering agent, to maintain the soil pH. The sample was filtered and the solid residue washed with fresh reagent. The filtrate was then used directly for FAAS.

Removing inorganic analytes from organic matrices

Ashing destroys the organic matrix and is normally done when the sample consists largely of organic matter such as soils or plants. There are two ways of ashing a sample – dry ashing and wet ashing.

Dry ashing. The sample is heated in a crucible in a muffle furnace. Ashings usually take between 12–15 hours at a temperature between 400–600 °C. The resulting ash is dissolved in dilute acid to give a solution of metal ions.

The main problems with dry ashing are:

- loss of volatile metals such as arsenic; and

- metals that are insoluble in the acid remain in the crucible after ashing.

Wet ashing. The problems associated with dry ashing can be overcome by wet ashing. In this process the sample is heated with oxidising agents – concentrated nitric acid followed by chloric(VII) acid – to break down any organic matter. Because strong oxidising agents are used, wet ashing can be done at lower temperatures than dry ashing. However, there is a risk of contamination from impurities in the acids. This is particularly important in trace analysis. Only small sample sizes can be ashed because chloric(VII) acid can detonate if the sample dries out.

Extracting organic analytes

The fields in which organic materials are normally analysed are clinical samples, environmental pollutants, for quality control in agriculture, medical products, petrochemicals, for trading standards in foods and drinks, for drug analysis in custom controls and environmental monitoring to meet health and safety standards. There is strict European legislation on the permitted levels of pesticides, phthalates, polychlorinated biphenols (PCBs), dioxins and mycotoxins.

Two basic methods for extracting organic solutes are shaking the sample with a solvent and filtering, or refluxing to improve solubility.

Soxhlet extraction (Box 2) is used for this type of extraction when the components can be prepared for analysis by liquid or gas chromatography, and this extraction method is widely used in industry.

RS•C

Soxhlet extraction Box 2

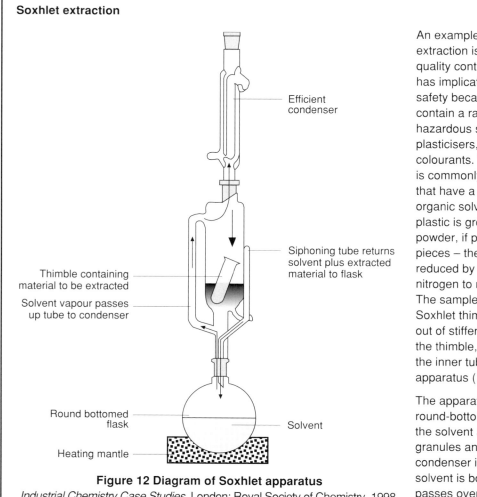

Efficient condenser

Siphoning tube returns solvent plus extracted material to flask

Thimble containing material to be extracted

Solvent vapour passes up tube to condenser

Round bottomed flask

Solvent

Heating mantle

Figure 12 Diagram of Soxhlet apparatus
Industrial Chemistry Case Studies, London: Royal Society of Chemistry, 1998.

An example where this type of extraction is used is in the quality control of plastics. This has implications for health and safety because plastics may contain a range of potentially hazardous substances such as plasticisers, antioxidants and colourants. The Soxhlet method is commonly used for solutes that have a low solubility in most organic solvents. The sample plastic is ground down into a powder, if possible, or into small pieces – the sample size can be reduced by freezing it in liquid nitrogen to make it more brittle. The sample is placed in the Soxhlet thimble, which is made out of stiffened filter paper, and the thimble, in turn, is fixed in the inner tube of the Soxhlet apparatus (*Figure 12*).

The apparatus is fitted into a round-bottomed flask containing the solvent and anti-bumping granules and then a reflux condenser is connected. As the solvent is boiled, the vapour passes over the thimble, condenses and drips back into the sample. When the solvent reaches the top of the thimble it is siphoned back into the flask, removing some of the solute from the sample. The process is repeated, extracting progressively more solute. However, the process is time-consuming – an efficient extraction requires about eight hours.

Trace metals

Many analyses require the determination of metals that are present in trace quantities. It is difficult to put an exact figure on the divide between trace and non-trace quantities. A metal may be present in reasonably high concentrations in a sample but may have had its concentration significantly lowered by the time it has been prepared for analysis. Some techniques may have detection limits, which are very low for some metals but higher for others. In general, concentrations of metals in ppm, mg kg^{-1}, or µg g^{-1}, approach trace levels.

Trace analysis is particularly important in environmental monitoring. Low concentrations of metals such as arsenic can be toxic. The cooling systems of nuclear power stations must be monitored for the slightest evidence of corrosion. Regular monitoring of the research and manufacturing environments of the semiconductor industry has to be done to check for trace contaminants. There are now strict regulations on the levels of toxic metals in children's toys, which require trace level determinations.

RS•C

In some cases the levels are so low that the sample containers or the blank reagents may have higher concentrations of the analyte than the sample. One solution to this problem is to use a technique that is extremely sensitive and has very low detection limits. However, there is usually a price to pay in these techniques.

Inductively coupled plasma-mass spectrometry (ICP–MS) has very low detection limits for most metals, down to parts per trillion (ppt). However, in many industrial contexts the analyte such as detergents, foods, ceramics, excipients (a substance mixed with a medicine to give it consistence, or used as a vehicle for its administration) in pharmaceuticals, is accompanied by a difficult matrix. In samples from medical laboratories, the matrix could be blood, slices of liver or muscle. The nickel skimmer cones at the interface of ICP–MS instruments are easily blocked by small particles, which means the matrix has to be removed or multiple dilutions of the sample have to be made until the detection limits are the same for less sensitive instruments. Inductively coupled plasma-mass spectrometry is best used for samples that are relatively matrix free – *eg* in the water industry.

Volatile elements

The analysis of volatile elements can best be done using hydride generation spectrometry. However, all methods require preconcentration. This is usually done by evaporation. For volatile elements preconcentration is done using distillation, after which the analyte is redissolved in a suitable solvent.

A suitable solvent can be used to preconcentrate analytes. This has a selective advantage because control of a factor like pH can result in extracting some analytes in preference to others. Solvent extraction usually involves the following steps:

- chelation;

- partition into an organic phase; and

- back extraction into an aqueous phase.

As an example, ammonium pyrrolidinedithiocarbamate (APDC) forms stable complexes with most transition metals – *eg* nickel and chromium. After chelation it can be partitioned from aqueous solvents into 1,1,2-trichloro-1,2,2-trifluoroethane by a separating funnel. Aqueous acids hydrolyse the complex, and the metals can then be back extracted into the aqueous phase. The analytes can then be determined by multi-element analysis, such as inductively coupled plasma atomic emission spectroscopy (ICP–AES).

Ion exchange

This process is described in *Modern Chemical Techniques,* London: Royal Society of Chemistry, 1992. As an example, chelating ion-exchange resins are used to separate trace metals from solutions containing a high concentration of alkali metal ions. Complexes are formed with the metal ions. These can later be recovered from the column by rinsing with acid.

Analytical accuracy

A number of factors are important when deciding on a particular analytical technique.

Limits of Detection (LoD)
This is the lowest concentration at which the analyte can be unequivocally identified. Practically, the LoD should be at least one-tenth of the concentration to be measured.

RS•C

For example, the legal limit for lead in tap water is 50 ppb (0.05 mg dm^{-3}). Any method used to determine lead in tap water for monitoring purposes should be capable of measuring lead in concentrations of 5 ppb (0.005 mg dm^{-3}).

Accuracy

This is the closeness of a result, or the mean of a set of results, to the true or accepted value. The true value can never be known exactly therefore the accuracy of a result cannot be known exactly.

The accuracy of a result is important in instances where the conviction of a person, such as in DNA fingerprinting, or the prosecution of a company may depend on it. Where the levels of toxic metals in an artifact are near the legal limit, accuracy is very important to determine that the levels are not exceeded. Accurate analysis is also important for checking purity – *eg* characterising gold. However, if the concentrations of lead are found to be around 1–2 ppb in drinking water then the accuracy of the results are not so important because they are well below the legal limit of 50 ppb.

If a sample has to be successively diluted this affects the accuracy of the results.

Precision

This is the closeness of a series of replicate results to each other. Precision is not the same as accuracy, because measurements can be precise but inaccurate.

Again, precision is only important where the results are close to the margins.

Interferences

Any analytical method must be capable of distinguishing between the analyte and other species present in the matrix. Usually, some separation process needs to be done to remove the analyte from the matrix.

Matrix viscosity may prevent efficient nebulisation of an analyte – *eg* in atomic absorption spectroscopy.

Isobaric interference where species have the same m/z ratio – *eg* ArO^+ and Fe^+ in inductively coupled plasma – mass spectrometry.

Spectral interference where emission wavelengths of two metals are very close, *eg* titanium masking chromium in inductively coupled plasma – atomic emission spectroscopy, similar 2θ values as in X-ray fluorescence spectrometry, or in electrophoresis when two proteins have identical values of pI.

Other components in the matrix may suppress atomisation – *eg* calcium phosphate in flame atomic absorption spectrometry.

Ionisation interference where ions may be formed rather than atoms – *eg* alkali metals in flame atomic absorption spectrometry.

Calibration

Calibration involves making a comparison of a measured quantity against a reference value. In a spectrometer, a standard reference material is used to measure the response of the spectrometer to the known concentration of the standard. Standards can be taken at different values and a calibration graph is drawn. The response of the spectrometer to a substance with an unknown concentration can then be made. Standards may serve different purposes and be graded in different ways. For example, a standard may be specific to a particular instrument in a laboratory or to a specific series of measurements, and have no applicability beyond these. Alternatively, an

RS•C

internationally recognised Certified Reference Material will have been validated by an internationally recognised procedure, and be applicable to a wide range of instruments.

Many calibrations are matrix dependent. For example, if chromium dissolved in phosphoric acid is being measured by ICP–OES then the standard should not only have an accurately known concentration of chromium, but the matrix should match by having the chromium dissolved in phosphoric acid. In XRF, solid standards are made for matrix matching. As an example, measuring the amount of manganese in steel involves using a standard with a known amount of manganese in an alloy with a composition as close as possible to that of the steel being tested.

Blanks

Blanks are used to check for contamination of reagents, for contaminants in the general laboratory atmosphere, or for the appearance of an interfering substance contained within the matrix of the analyte.

Blanks contain all the reagents with the exception of the analyte. The measured concentration of the analyte should be the difference between that recorded by the sample and that recorded by the blank.

Uncertainty

This characterises the range of values within which the value of the quantity being measured is expected to lie. It is an indicator of quality in chemical measurements.

RS•C

Bibliography

R. Beaty and J. Kerber, *Concepts, instrumentation and techniques in atomic absorption spectrophotometry*, Perkin-Elmer, 1993.

L. Lajunen, *Spectrochemical analysis by atomic absorption and emission*, Cambridge: Royal Society of Chemistry, 1992.

E. Metcalfe, *Atomic absorption and emission spectroscopy*, John Wiley, 1987.

S. Bradbury, *An introduction to the optical microscope*, Oxford Science Publications, 1989.

E. Gravè, *Using the microscope*, Dover Publications Inc, 1984.

D. Simpson and W. Simpson, *An introduction to applications of light microscopy in analysis*, Cambridge: Royal Society of Chemistry, 1988.

L. Smart and E. Moore, *Solid State Chemistry*, London: Chapman & Hall, 1992.

C. Whiston, *X-ray Methods*, Chichester: John Wiley and Sons on behalf of ACOL, 1987.

S.E. Dann, *Reactions and Characterization of Solids*, London: Royal Society of Chemistry, 2000.

R. Anderson, *Sample Pretreatment and Separation*, Chichester: John Wiley and Sons on behalf of ACOL, 1987.

E. Prichard, *Quality in the AnalyticalChemistry Laboratory*, Chichester: John Wiley and Sons on behalf of ACOL, 1995.

E. Prichard, G.M. Mackay, and J. Points, (eds) *Trace analysis: A structured approach to obtaining reliable results,* London: Royal Society of Chemistry for LGC, 1996.

B. Woodget, D. Cooper, *Samples and Standards*, Chichester: John Wiley and Sons on behalf of ACOL, 1987.

B.D. Hames and D. Rickwood, *Gel electrophoresis of proteins*, IRL, 1990.

M. Melvin, *Electrophoresis*, Wiley, 1987.

P. Jandik and G. Bonn, *Capillary electrophoresis of small molecules and ions*, VCH, 1993.

Y. Li, Capillary electrophoresis: *Principles, practice and applications*, Amsterdam: Elsevier, 1992.

R. Jenkins, *X-ray fluorescence spectrometry*, Chichester: John Wiley, 1988.

C. Vandecasteele and C.B. Block, *Modern methods for trace element determination*, Chichester: Wiley, 1993.